AI Mastery for Pickleball Business Owners

Actionable Strategies to Automate, Scale, and Profit

First edition

AI Mastery for Pickleball Business Owners Actionable Strategies to Automate, Scale, and Profit

By Elena Saris

CeolWaves

2025

UNITED STATES OF AMERICA

Table of Contents

FOREWORD

Harnessing AI for a Competitive Edge in the Pickleball Industry

The pickleball phenomenon is undeniable. This rapidly expanding sport has created a vibrant ecosystem of businesses, all vying for success in a dynamic and increasingly competitive market. For entrepreneurs and business leaders in the pickleball sector, the question is no longer just about participating in this growth, but about strategically navigating it to achieve sustainable success. The answer, in significant part, lies in the intelligent adoption of emerging technologies.

At the forefront of this technological evolution is Artificial Intelligence (AI). Once thought of as a complex tool reserved for universities and large corporations, AI has become remarkably accessible, offering a suite of powerful capabilities that can provide a distinct competitive advantage to businesses of all scales. For the pickleball industry, AI represents a pivotal opportunity to enhance marketing effectiveness, deepen customer engagement, and optimize operational efficiency.

"AI Mastery for Pickleball Business Owners" is designed as a practical guide for business professionals like you. It aims to demystify AI and demonstrate its tangible applications within the specific context of the pickleball market. This book will equip you with the knowledge and strategies to integrate AI into your operations, whether your goal is to sell gear or B2B services, expand your club membership, fill your retreats, promote your coaching, or launch innovative pickleball-related software.

The core premise is that AI can amplify your business acumen. It can provide rapid market insights, assist in crafting targeted and compelling marketing content, and automate routine tasks, thereby freeing your valuable time to focus on high-level strategy, innovation, and building strong customer relationships. This is not about replacing human expertise but augmenting it with data-driven insights and enhanced capabilities.

This guide will explore how AI can be applied to both Business-to-Consumer (B2C) and Business-to-Business (B2B) models within the pickleball industry. It offers actionable advice and specific examples suitable for those new to AI, as well as for those looking to refine their existing AI strategies. We will explore a range of AI tools, from widely used platforms for content creation and data analysis to more specialized applications, ensuring you understand how to select and implement solutions that best fit your business needs.

The integration of AI is rapidly becoming a standard for innovative businesses. For the pickleball community, this presents a significant opportunity to lead, adapt, and thrive. This book serves as your roadmap to understanding and utilizing AI, enabling you to build a more resilient, efficient, and successful pickleball enterprise. The tools are available; the time to leverage them is now.

NOTE TO READERS

Beyond the Book: Resources for Implementation

While this book provides you with everything needed to implement AI marketing strategies for your pickleball business, we recognize that some readers prefer to focus their time on what they do best while leveraging expert support for specialized tasks.

To that end, consider these two complementary resources:

Right On Time Data: Done-For-You Research Services

For business owners who want professional-grade research without the time investment, we can create any of the intelligence reports described in this book specifically for your business:

- Market Research Analysis - Industry trends and opportunities
- Avatar Research Profiles - Your ideal customer characteristics
- Intent-Based Audience Mapping - Prospects ready to buy
- Competitive Intelligence Reports - Market positioning insights
- Partnership Opportunity Analysis - Strategic collaboration prospects

Recreational Pickleball Players Association (RPPA) – Vendors Directory

Beyond research and strategy, success in the pickleball industry requires authentic community engagement. For vendors seeking direct access to engaged pickleball players and fellow business owners:

• Create a branded profile and showcase products with images and video
• Offer exclusive VIP deals to engaged, purchase-ready players
• Get multichannel exposure across our app, newsletter, and our website
• Offer VIP club membership and directly DM or broadcast to those who opt in
• Build authentic relationships with everyday players

These resources work together to support your success - combining strategic intelligence with community engagement for maximum impact.

Scan this QR Code to Learn More about one or both:

INTRODUCTION

Applying AI for Strategic Growth in Your Pickleball Business

This guide is for business professionals engaged in the dynamic and rapidly expanding pickleball industry. If you are seeking innovative strategies to enhance your market presence and operational efficiency, "AI Mastery for Pickleball Business Owners" offers a practical framework for integrating Artificial Intelligence into your core business functions.

The Expanding Pickleball Market: Opportunities and Challenges

The pickleball sector is experiencing significant growth, with increasing player participation, demand for facilities and equipment, and a burgeoning ecosystem of related services. This expansion presents considerable opportunities. However, it also intensifies competition, necessitating sophisticated approaches to marketing, customer engagement, and business development. Effectively utilizing technology, particularly AI, can provide a critical advantage in this evolving landscape.

Understanding Artificial Intelligence in a Business Context

Artificial Intelligence, at its core, involves computational systems performing tasks that traditionally require human intellect, such as learning from data, solving problems, and making informed decisions. Many businesses already interact with AI through various applications, including automated customer service, data analysis tools, and content generation platforms.

This book will concentrate on practical AI applications relevant to pickleball businesses, including:

Large Language Models (LLMs): Systems like ChatGPT and Claude, which process and generate human-like text for content creation, communication, and strategic brainstorming.

Generative AI (GenAI): AI capable of creating novel content, including text, images, and other media, based on user prompts. Tools such as Canva AI, DALL-E, and Midjourney fall into this category.

AI-Powered Analytics: Applications that analyze business data to identify trends, predict outcomes, and provide actionable insights for marketing and operational strategies. Tools such as Gemini (formerly Bard), NotebookLM and Microsoft Power BI with Copilot.

Autonomous AI Agents: Advanced AI systems, exemplified by Manus, designed to manage complex, multi-step workflows. These agents can integrate various tools and data sources to automate tasks like market research, report generation, and targeted outreach campaigns, operating with a degree of autonomy based on strategic directives.

Our objective is to clarify these technologies and demonstrate their practical implementation, enabling you to leverage AI regardless of your current level of technical expertise.

The Strategic Advantages of AI in Pickleball Marketing

Integrating AI into your marketing and business operations offers several key benefits:

Increased Efficiency: Automate routine and time-consuming tasks, allowing your team to focus on strategic initiatives and high-value customer interactions.

Enhanced Customer Insights: Utilize AI to analyze customer data and feedback, leading to a deeper understanding of their needs, preferences, and behaviors.

Scalable Personalization: Deliver tailored marketing messages and customer experiences at scale, improving engagement and loyalty.

Optimized Content Creation: Streamline the development of high-quality, relevant content for various marketing channels.

Data-Informed Decision-Making: Base your strategic decisions on robust data analysis and predictive insights provided by AI tools.

High-Impact Market Research & Competitive Intelligence: New Generative AI tools can generate full industry reports, analyze customer behavior across platforms, surface hidden trends, and map out competitor strategies in minutes.

Illustrative Scenario: Enhancing a Pickleball Tournament Promotion

Traditional Approach: Manual creation of promotional materials, broad distribution of generic announcements, and limited ability to track engagement effectiveness across different channels.

AI-Enhanced Approach: Employ LLMs to draft targeted promotional copy for different player segments. Use Generative AI to create professional-quality visuals for social media and digital ads. Utilize AI-powered email marketing tools to personalize invitations and reminders. Task an AI agent like Manus to monitor social media for mentions of local tournaments and identify potential attendees or influencers for targeted outreach, while also compiling engagement analytics from various platforms into a consolidated report.

Target Audience: Professionals Across the Pickleball Sector

This guide is intended for individuals and organizations aiming to achieve growth and competitive differentiation in the pickleball market. This includes:

- **Pickleball Entrepreneurs and Innovators**
- **Club Owners and Facility Operators**
- **Software and App Developers serving the pickleball market**
- **Coaches, Trainers, and Event Organizers**
- **Equipment Manufacturers, Retailers, and E-commerce Businesses**

The principles and strategies discussed are applicable to both B2C and B2B business models within the pickleball industry.

Navigating This Guide: A Roadmap to AI Implementation

This book is structured to provide a clear and actionable path to AI integration.

For those newer to AI: The initial sections establish a foundational understanding of AI in the context of market analysis and strategic planning. Subsequent chapters introduce core AI applications with clear explanations and practical examples.

For those with existing AI experience: The guide offers advanced strategies, including the integration of multiple AI tools, sophisticated prompt engineering techniques, and the utilization of autonomous agents like Manus for complex workflow automation. The detailed examples and tool discussions aim to provide new perspectives for experienced users.

Focus on Practical Implementation: Each chapter emphasizes real-world application, featuring:

Relevant AI Tool Spotlights: Highlighting effective and accessible tools for pickleball businesses, with an understanding of their specific strengths.

Adaptable Prompt Frameworks: Providing examples of prompts that can be customized for various business scenarios to achieve desired outcomes.

Actionable Strategies: Detailing how to implement AI solutions in a structured manner.

Consider this book a strategic resource. You may read it sequentially or focus on chapters addressing your most immediate business challenges. The key is to experiment with the concepts, adapt them to your unique business context, and iterate based on your results. The field of AI is characterized by rapid evolution, and your proficiency will grow with application and experience.

Your initiative to explore AI for your pickleball business is a step towards future success. Let us begin.

Note: This book was created mainly with the help of Manus AI, a tool I trained and directed to reflect my thoughts, expertise, and intent. I strategically combined multiple AI platforms (including ChatGPT, Perplexity, NotebookLM, Gemini and Flowith) to ensure comprehensive accuracy and cross-verification of all content—the result reflects my own expertise and strategic thinking, while the process itself serves as a real-world example of what's possible when human insight meets AI efficiency.

Throughout this book I will refer to Manus, since as of this writing it is the most accessible Agentic AI Agent. Any time you see "Prompt for Manus" understand this to mean this is a prompt for you to use with any of your current Agentic AI Agents or platforms.

CHAPTER 1

Understanding Your Pickleball Market with AI

In the dynamic and rapidly growing pickleball sector, a thorough understanding of your market is essential for sustainable business growth. Before effectively marketing your products, services, or facilities, it is crucial to develop a clear perspective on the industry landscape. This includes identifying potential customers, understanding their needs and preferences, recognizing your competitors and their strategies, and staying informed about prevailing industry trends. Traditionally, gathering this market intelligence required considerable time and resources. Today, Artificial Intelligence provides a range of powerful tools that can make market research more accessible, offering deep, actionable insights with greater speed and precision.

This chapter will guide you in using AI to establish a solid foundation for your marketing strategy by comprehensively understanding your pickleball market. We will explore how AI can assist in identifying your specific business niche, analyzing market trends, conducting thorough competitor analysis, and extracting valuable insights from online community discussions.

Identifying Your Niche in the Pickleball Ecosystem (B2C vs. B2B)

The pickleball market is diverse, featuring a wide array of businesses that cater to various segments. Your enterprise might be a Business-to-Consumer (B2C) operation, directly serving players with equipment, apparel, coaching, or club memberships. Alternatively, you could operate as a Business-to-Business (B2B) entity, providing services such as court construction, club management software, or wholesale equipment to other businesses within the pickleball industry. Many businesses also adopt a hybrid model, serving both consumers and other businesses.

A clear definition of your niche is a critical first step. AI can aid in this process by helping you brainstorm and assess potential niches based on factors like market saturation, existing demand, and your company's unique strengths. For example, rather than broadly targeting all pickleball players, AI will help you identify an underserved sub-segment, such as competitive tournament players over the age of 50, or beginners interested in environmentally sustainable pickleball gear.

To illustrate how AI can assist in brainstorming niche opportunities, consider using a Large Language Model (LLM) like ChatGPT or Claude.

Prompt Example (for ChatGPT/Claude):

"Assume the role of an experienced market analyst specializing in the sports industry. My objective is to identify underserved or high-potential niche markets within the U.S. pickleball ecosystem, considering both B2C and B2B opportunities. Generate 10 distinct niche ideas. For each idea, provide: 1. A concise description of the niche. 2. The primary target

audience (consumer or business type). 3. Potential unique selling propositions (USPs) for a business operating in this niche. 4. Key considerations or potential challenges associated with this niche."

The AI's output can provide a structured list of potential niches, complete with initial considerations regarding target audiences and USPs, offering a valuable starting point for more detailed investigation.

AI Tools for Market Trend Analysis

Staying informed about market trends is vital. Is there an increasing demand for indoor courts in a specific region? Are particular paddle technologies gaining traction? Are player demographics evolving? AI tools can help answer these questions by analyzing extensive data from various sources.

One such tool is Perplexity AI, which excels at real-time, web-connected research and can provide sourced answers to inquiries about current trends. It synthesizes information from recent articles, reports, and discussions, offering a quick overview of current market dynamics.

Tool Spotlight: Perplexity AI

* **Strengths:** Provides up-to-date information with direct source citations, making it effective for quick research, fact-checking, and understanding current discussions. Its conversational interface allows for follow-up questions to explore topics in greater depth.

* **Ideal Use Cases for Pickleball Businesses:** Researching new equipment trends, understanding regional demand for pickleball facilities, tracking competitor announcements, and fact-checking claims before publishing content.

Prompt Example (for Perplexity AI):

"Summarize the top market trends in the U.S. pickleball paddle industry from the past six months — especially around materials (e.g., carbon fiber vs. fiberglass), price trends, and what players are saying about performance and preferences. Include links to recent articles, retail data, and player reviews if available."

Another approach involves using Google Trends in conjunction with AI analysis. While Google Trends offers valuable data on search interest over time, combining this with AI can unlock deeper insights. You can export Google Trends data and utilize AI tools, including capable LLMs, to analyze patterns, correlate trends with specific events, or forecast potential future interest based on historical data. An AI agent like could be tasked to periodically retrieve Google Trends data for specific pickleball-related keywords, analyze it for significant shifts, and incorporate these findings into a regular market update report.

Competitor Analysis with AI: Uncovering Strengths, Weaknesses, and Opportunities

Understanding your competition is as crucial as understanding your customers. AI can automate and enhance many aspects of competitor analysis, from identifying your competitors to dissecting their strategies.

AI-powered searches can help identify direct and indirect competitors in your specific niche and geographic area. Furthermore, AI tools can scan competitor websites, social media profiles, and customer reviews to extract information about their product offerings, pricing, marketing messages, customer service quality, and overall brand perception.

Prompt Example (for ChatGPT/Claude, after providing URLs or text from a competitor's site):

"Analyze the provided website content for [Competitor's Pickleball Club Name]. Identify: 1. Their primary target audience as suggested by the website's language and offerings. 2. Their key services and amenities. 3. Their main selling propositions or differentiators. 4. The overall tone and brand voice of their communications. 5. Any apparent weaknesses or areas where my new pickleball club could offer a superior alternative."

For ongoing competitor monitoring, an autonomous AI agent such as can be particularly useful. You could configure a task for to regularly monitor the websites and social media feeds of your key competitors.

Example Task for :

"On a weekly basis, monitor the websites ([URL1], [URL2], [URL3]) and Instagram profiles ([Profile1], [Profile2], [Profile3]) of my top three local pickleball competitor clubs. Track and report any of the following changes or new information:

* New membership offers or pricing adjustments.

* Announcements of new programs, clinics, or events.

* Significant new facility upgrades or additions.

* Special promotions or discounts being offered.

* Highly engaging social media posts (based on likes and comments). Compile these findings into a concise weekly report, highlighting the most strategically important updates."

This type of automated monitoring ensures you remain informed about your competitors' activities without dedicating extensive hours to manual checks.

Listening to the Community: AI for Social Listening and Insights

The pickleball community is highly active online across various platforms, including forums, Facebook groups, Instagram, X (formerly Twitter), and Reddit. These platforms are rich sources of information regarding player opinions, unmet needs, emerging trends, and sentiment towards brands and products. Manually processing this volume of information is impractical, but AI can perform this task efficiently.

Specialized AI tools are designed for social listening. They can track mentions of keywords (such as your brand, your competitors, or specific pickleball topics), analyze the sentiment (positive, negative, neutral) of these mentions, and identify key influencers or trending discussions. Brandwatch is an example of a social listening tool.

Even without specialized tools, LLMs can be leveraged to analyze text data extracted from online communities (always ensuring respect for platform terms of service and user privacy). For instance, you could gather public comments from a pickleball forum discussion about a new paddle type and use an LLM to summarize the main points of praise and criticism.

Prompt Example (for ChatGPT/Claude, after providing a collection of anonymized comments):

"Analyze the following set of public comments from a pickleball forum discussing the new 'BrandX CarbonGlide' paddle. Identify:

1.The top three most frequently mentioned positive attributes of the paddle.

2.The top three most frequently mentioned negative attributes or concerns.

3.The overall sentiment expressed (positive, negative, mixed).

4.Any suggestions for improvement mentioned by the users.

5.Emerging themes or topics of discussion related to this paddle."

An AI agent like Manus could also be tasked to monitor specific online pickleball communities or forums for mentions of your brand or relevant keywords. It could flag urgent customer service issues, identify frequently asked questions that could inform your FAQ page, or alert you to emerging discussions about new product desires. There are advanced ways to automate these continuous tasks and while those are beyond the scope of this book, the prompts are designed to permit the least amount of manual input.

Prompt Example (for to monitor these forums and mentions):

"Set up continuous monitoring of public online communities, forums, and review sites where pickleball players discuss products. Track all mentions of the following keywords: 'BrandX', 'CarbonGlide', 'carbon fiber paddle', and 'lightweight pickleball paddle.

Each week I will ask you to generate a summary that includes:

-The top 5 frequently mentioned features (positive or negative) of the BrandX CarbonGlide paddle.

-Any urgent or recurring customer service issues or complaints.

-A list of frequently asked questions that could be addressed in our support materials or website.

-Suggestions or requests for product improvements.

-Emerging competitor mentions or comparisons.

-A sentiment breakdown with volume trends.

Include direct quotes where helpful and link to original posts when available. Output as a weekly report in plain language, designed for marketing and product teams."

By systematically applying AI to understand your market niche, track trends, analyze competitors, and listen to community conversations, you build a robust intelligence capability for your pickleball business. These insights will inform not only your marketing messages but also your product development, service offerings, and overall business strategy, ensuring you are well-positioned to effectively serve the needs of the dynamic pickleball market. The next chapter will build upon this foundation by focusing on how AI can help you define and understand your ideal customer in even greater detail.

[Note: As I am going to press on this edition, Manus has just announced that they will be able to self-schedule these types of tasks so that reports will be provided to you on a schedule without the need for separate prompts.]

CHAPTER 2

Defining Your Ideal Pickleball Customer/Client with AI

Following our exploration of the broader pickleball market landscape using AI, this chapter focuses on the most critical component of any successful business: the customer. A precise understanding of your target audience – encompassing their demographics, psychographics, behaviors, needs, and challenges – is essential. A vaguely defined audience can lead to ineffective marketing messages, misallocated resources, and missed opportunities. Artificial Intelligence offers robust capabilities to move beyond generic customer profiles and develop nuanced, data-driven understandings of your ideal pickleball customers (for B2C operations) or clients (for B2B operations).

This chapter examines how AI can assist in creating detailed customer personas, segmenting your audience for targeted communication, and uncovering the true needs and motivations that drive their engagement with the pickleball industry. These insights are vital for crafting marketing campaigns that resonate, developing products that meet demand, and offering services that cultivate lasting loyalty.

Creating Detailed Customer Personas with AI

A customer persona is a semi-fictional, research-based representation of your ideal customer. It extends beyond simple demographics to include goals, motivations, challenges, concerns, and decision-making processes. Well-crafted personas enable businesses to empathize with their audience and tailor marketing efforts more effectively. AI, particularly Large Language Models (LLMs), can significantly accelerate and enrich the persona creation process.

To illustrate, you can provide an LLM like ChatGPT or Claude with existing customer data (anonymized and aggregated), survey results, or even general knowledge about pickleball player archetypes to generate comprehensive personas.

Prompt Example (for ChatGPT/Claude):

"Act as a marketing strategist for a new upscale pickleball club situated in a suburban area with a diverse population including young professionals and active retirees. Our objective is to develop detailed customer personas to inform our marketing strategies. Based on general knowledge of these demographic groups and their potential interest in pickleball, generate two distinct customer personas:

Persona 1: Young Professional (Age 28-40) Persona 2: Active Retiree (Age 60-75)

For each persona, include: **1.** A representative name and a brief physical description (e.g., "Energetic Emily, mid-30s, athletic build"). **2.** Demographics: Age range, typical occupation, general income bracket, family status. **3.** Pickleball Experience Level: (e.g., Beginner, Intermediate, Advanced, New to the sport). **4.** Goals & Motivations for Playing

Pickleball: (e.g., Fitness, social interaction, competition, stress relief, skill development). **5.** Challenges & Pain Points Related to Pickleball: (e.g., Difficulty finding court time, lack of suitable playing partners, risk of injury, cost of equipment/ membership, perceived intimidation for beginners). **6.** Information Sources & Channels They Trust: (e.g., Social media platforms, local community forums, word-of-mouth referrals, sports publications, specific websites). **7.** Key Marketing Messages Likely to Resonate: (e.g., "Connect and compete in a vibrant community," "Stay active and social through pickleball," "Learn pickleball in a supportive and welcoming environment"). **8.** Potential Objections to Joining Our Upscale Club: (e.g., Membership fees, perceived exclusivity, commute time).

Ensure each persona is vivid, relatable, and grounded in plausible characteristics."

The AI will generate detailed profiles that can then be validated and refined with your specific business data. These personas become invaluable reference points for your entire team when making marketing and operational decisions.

AI-Driven Segmentation: Targeting the Right Audience Segments

Not all pickleball players or businesses are identical. Market segmentation involves dividing your broader market into smaller, more manageable groups that share common characteristics. This practice allows for more targeted and effective marketing initiatives. AI can analyze your customer data (e.g., from your CRM system, email marketing list, or sales records) to identify meaningful segments based on various criteria:

- **Demographic Segmentation:** Based on age, gender, location, income, occupation.
- **Psychographic Segmentation:** Based on lifestyle, values, interests, personality traits (e.g., distinguishing between competitive players and social players, or early adopters of new equipment versus value-conscious buyers).
- **Behavioral Segmentation:** Based on purchasing habits, product usage patterns, engagement with your brand, skill level, and frequency of play.
- **Needs-Based Segmentation:** Grouping customers based on the specific benefits they seek from products or services (e.g., players seeking high-performance equipment versus those prioritizing comfort and affordability).

If you possess a customer database, AI tools can often connect to it (or analyze an exported version) to identify patterns and suggest segments you might not have previously considered.

Even if you do not possess a customer database, AI can still support effective segmentation. Generative AI tools like ChatGPT or Claude can create realistic, research-based customer personas by drawing on general knowledge of pickleball demographics and consumer behavior. These hypothetical segments can help guide your early marketing decisions. Additionally, AI-powered research tools such as Perplexity or can analyze public sources—like online forums, product reviews, and social media discussions—to identify emerging trends, common preferences, and pain points. This external data can help you infer likely customer segments even before you've made your first sale.

Prompt Example (Conceptual, for an AI analysis tool or LLM with data analysis capabilities):

"Analyze the provided customer dataset for our online pickleball equipment store. The dataset includes purchase history (items bought, purchase frequency, average order value), email engagement metrics (open rates, click-through rates), and basic demographics (age, location, where available). Identify at least three distinct customer segments based on purchasing behavior and engagement levels. For each segment, describe: **1.** Key characteristics and behaviors. **2.** Potential marketing approaches tailored to this segment.

For instance, one segment might be identified as 'High-Value Enthusiasts' who frequently purchase premium paddles and consistently engage with email content, contrasted with 'Occasional Players' who infrequently buy entry-level equipment."

Understanding these segments enables you to tailor email campaigns, advertising targeting, and content creation more precisely. For example, a pickleball apparel brand might develop different marketing messages for a segment of fashion-conscious younger players compared to a segment of older players who prioritize comfort and sun protection.

Understanding Customer Needs and Pain Points with AI Surveys and Feedback Analysis

Directly soliciting feedback from your customers (or potential customers) about their needs and challenges provides invaluable information. AI can assist in both creating effective surveys and, importantly, in analyzing the responses – particularly open-ended qualitative feedback, which can be time-consuming to process manually.

LLMs can help draft survey questions that are clear, unbiased, and designed to elicit the specific information you require.

Prompt Example (for ChatGPT/Claude):

"I'm developing a member survey for a suburban pickleball club to assess satisfaction and gather ideas for improvement. Please create five **open-ended** and five **multiple-choice** questions that cover: court availability, facility quality, program variety (e.g., leagues, clinics), social engagement opportunities, and communication effectiveness.

Ensure the questions are **neutral and inclusive**, designed to prompt honest feedback from players of all skill levels. Where relevant, suggest multiple-choice formats (e.g., Likert scale, yes/no, or ranking).

At the end, include 2 optional **demographic or play-level questions** to help segment responses during analysis."

AI excels in feedback analysis. You can input open-ended survey responses, customer service transcripts, online reviews, or social media comments into AI tools to identify recurring themes, overall sentiment, and key pain points.

Tool Spotlight: NotebookLM - AI-Powered Survey Analysis

***Strengths:** NotebookLM allows you to upload your own documents — including survey responses, spreadsheets, and reports — and then query them using natural language. It excels at summarizing open-ended feedback, identifying recurring themes, surfacing sentiment, and highlighting actionable suggestions, all without requiring manual tagging or data cleaning. It's especially useful for making qualitative survey results instantly usable for small teams.

***Ideal Use Cases for Data:**

- Summarizing post-event or post-program feedback
- Understanding key reasons for membership churn or declining engagement
- Surfacing patterns in open-ended survey responses about facility quality or communication
- Identifying recurring feature requests or pain points for program improvements

Prompt Example (for NotebookLM, after providing a list of anonymized customer complaints):

"Analyze the following customer complaints collected from our pickleball club members. For each of the five key operational areas — court availability, facility cleanliness/ maintenance, program variety (e.g., leagues or clinics), staff communication, and overall member experience — do the following:

1) Identify the most commonly reported issues.

2) Summarize sentiment (positive, negative, mixed) related to each area.

3) Highlight any recurring suggestions or requests for improvement.

4) Flag any complaints that may represent urgent or reputational risk.

Present your findings as a structured summary with bullet points under each category."

For a continuous feedback mechanism, an autonomous agent like Manus could be tasked to periodically gather feedback from various channels (e.g., a dedicated feedback email address, specific social media threads, or online review sites). It could then perform an initial analysis, categorize the feedback, identify urgent issues requiring immediate attention, and compile a summary report for your review. This creates an ongoing feedback loop, helping your business stay responsive to evolving customer needs.

Example Task for Manus:

"Monitor the following sources for customer feedback regarding our pickleball coaching services: [Feedback Email Address], [Specific Facebook Group Thread URL], [Google My Business Review Page URL]. Once a week, compile all new feedback. Analyze this feedback to identify: *Positive comments and testimonials (extract illustrative quotes). *Common themes in negative feedback or suggestions for improvement. *Any urgent issues that require immediate attention. Be prepared to summarize these findings in a weekly report, including representative quotes and an overview of customer sentiment."

By leveraging AI to create detailed personas, segment your audience effectively, and deeply understand customer needs and pain points, you build a customer-centric foundation for your pickleball business. This profound understanding will not only shape your marketing messages but also influence your product development, service offerings, and overall customer experience strategy. The more intimately you know your ideal customer, the better equipped you are to serve them, and AI is an increasingly indispensable tool in achieving that level of understanding. In the next chapter, we will explore how to translate this market and customer insight into a cohesive, AI-infused marketing strategy.

CHAPTER 3

Developing Your AI-Infused Pickleball Marketing Strategy

With a clearer understanding of the pickleball market landscape (Chapter 1) and a more defined picture of your ideal customer (Chapter 2), it is time to construct the strategic framework that will guide your marketing efforts. A well-defined strategy serves as your roadmap, ensuring that your actions are purposeful, your resources are allocated effectively, and your goals are achievable. Artificial Intelligence can play a significant role not only in informing this strategy through data analysis but also in helping you articulate, refine, and even automate parts of its execution.

This chapter focuses on how to develop a robust, AI-infused pickleball marketing strategy. We will cover setting SMART (Specific, Measurable, Achievable, Relevant, Time-bound) goals with AI assistance, integrating AI seamlessly into your existing or planned marketing funnel, budgeting for AI tools and resources, and navigating the crucial ethical considerations of using AI in your marketing practices.

Setting SMART Marketing Goals with AI Support

Effective marketing begins with clear objectives. The SMART framework is a widely accepted method for setting goals that are well-defined and trackable. AI can help you transform vague aspirations into concrete SMART goals.

- **Specific:** Clearly define what you want to achieve. Instead of a general goal like "increase club memberships," a specific goal might be "increase new adult club memberships by 15%."
- **Measurable:** Define how you will track progress and success. For the membership goal, this would involve tracking the number of new sign-ups.
- **Achievable:** Ensure your goals are realistic given your resources and market conditions. AI-driven market analysis, as discussed in Chapter 1, can help assess achievability.
- **Relevant:** Align your marketing goals with your overall business objectives. Does increasing adult memberships support your broader business growth plan?
- **Time-bound:** Set a deadline for achieving your goal, for example, "within the next six months."

If you have a general marketing objective, an LLM can help you break it down and refine it into a SMART goal. Be specific.

Prompt Example (for ChatGPT/Claude):

"I own a new online store that sells premium pickleball paddles. My general marketing goal is to increase sales and build brand awareness, but I need help turning that into specific SMART goals.

Act as a marketing consultant. Based on the fact that I'm targeting adult recreational pickleball players (ages 35–65), and

I currently average 100 website visitors per day with a small budget focused on online ads and content creation, help me create **three distinct SMART marketing goals for Q3**.

For each goal, clearly define the Specific, Measurable, Achievable, Relevant, and Time-bound elements. Format each SMART goal in bullet points for easy reference."

The AI can help you articulate specific targets for website traffic, conversion rates, social media engagement, or sales figures, along with realistic timelines and methods for measurement.

Integrating AI into Your Existing Marketing Funnel

The marketing funnel (or flywheel, depending on your preferred model) describes the customer's journey from initial awareness of your brand to becoming a loyal advocate. AI can enhance every stage of this journey:

Awareness: Making potential customers aware of your pickleball business and what you offer.

AI Applications: SEO optimization for your website content (using AI keyword research tools), AI-generated blog posts and articles, AI-powered social media content creation and ad targeting, AI tools for identifying relevant influencers.

Manus Example: You could task Manus to research trending pickleball topics and generate content outlines for your blog to improve organic search visibility.

Interest/Consideration: Engaging potential customers and encouraging them to learn more about your offerings as they evaluate their options.

AI Applications: Personalized email newsletters with AI-curated content, AI chatbots on your website to answer questions 24/7, AI-generated product comparison guides or case studies, targeted retargeting ads.

Manus Example: Manus could monitor website visitor behavior (anonymized) and trigger personalized follow-up email sequences based on the pages they visited, offering more information on specific paddles or club services.

Decision/Conversion: Persuading interested leads to take action – make a purchase, sign up for a membership, book a court, or request a demo.

AI Applications: AI-powered CRM for lead evaluation and prioritization, AI-generated sales email copy, personalized offers and discounts delivered via AI, AI tools for optimizing landing page conversions.

Manus Example: For a B2B pickleball software company, Manus could be tasked to research potential leads (clubs, facilities), draft personalized introductory emails highlighting how the software solves their specific pain points (gleaned from online reviews or their website), and schedule follow-up reminders for the sales team.

Loyalty/Advocacy: Turning customers into repeat buyers and enthusiastic advocates for your brand.

AI Applications: AI-driven customer service (chatbots, automated responses), personalized post-purchase follow-ups, AI-powered loyalty program recommendations, AI analysis of customer feedback to improve services, encouraging and collecting online reviews.

Manus Example: Manus could monitor social media for mentions of your brand and flag positive testimonials that can be shared or identify customers who frequently engage positively and suggest them for a loyalty spotlight or special offer.

Walkthrough: AI in a Pickleball Club's Funnel

Imagine a pickleball club using AI throughout its member acquisition funnel:

* **Awareness:** AI helps create SEO-optimized blog posts about "learning pickleball in [City]" and targets local Facebook ads (with AI-generated copy and visuals) to people interested in sports and fitness.

* **Interest:** An AI chatbot on the club's website answers questions about membership tiers and court availability. Visitors who provide their email receive an AI-generated newsletter with tips for beginners and member testimonials.

* **Decision:** Manus identifies website visitors who spent significant time on the membership page but did not sign up and triggers a personalized email from the club manager offering a free introductory clinic.

* **Loyalty:** After joining, members receive AI-personalized updates about events matching their playing level. Manus can monitor club feedback forms and online reviews, summarizing key themes for management to address, thus improving member satisfaction.

Budgeting for AI Tools and Resources

The cost of AI tools varies widely. Many powerful tools offer free tiers or affordable subscriptions suitable for small to medium-sized pickleball businesses, while others, especially specialized enterprise solutions or extensive use of premium APIs, can represent a more significant investment.

When budgeting for AI:

* **Start with Free/Freemium Tools:** Explore the capabilities of tools like ChatGPT's free version, Canva AI's free features, Perplexity's free tier, and the free offerings of various email marketing or social media platforms with AI enhancements. This allows you to understand their value before committing financially.

* **Prioritize Based on Need:** Identify your biggest marketing pain points or the areas where AI could have the most immediate impact. Invest in tools that address these critical needs first.

* **Consider the ROI:** Think about the potential return on investment. If an AI tool costs $50 per month but saves you 10 hours of manual work (valued at, say, $30 per hour), or helps you acquire five new members (each worth a certain amount in revenue), the investment is likely worthwhile.

* **Factor in Learning Time:** While many AI tools are user-friendly, there is still a learning curve. Allocate time for yourself or your team to learn how to use new tools effectively.

* **Scalability:** Choose tools that can scale with your business. As you grow, your AI needs may evolve.

* **Manus and Resource Optimization:** Consider how an autonomous agent like Manus might fit into your budget. While there is a cost associated with such advanced automation, Manus can potentially optimize the use of other resources (including your time and subscriptions to multiple other tools) by streamlining workflows, automating data collection that you might otherwise pay a human or another specialized tool to do, and improving overall operational

efficiency. Its value often lies in its ability to consolidate and automate tasks that would otherwise require multiple disparate tools or significant manual effort.

Ethical Considerations and Responsible AI Use in Marketing

As with any powerful technology, using AI in marketing comes with ethical responsibilities. Building trust with your customers is paramount, and transparent, responsible AI use is key.

Data Privacy: Be transparent about how you collect and use customer data with AI. Comply with all relevant data privacy regulations (e.g., GDPR, CCPA). Ensure data used to train or inform AI models is handled securely and, where possible, anonymized.

Bias in AI: AI models can inadvertently perpetuate or even amplify existing biases present in their training data. This could lead to unfair or discriminatory marketing practices. Be mindful of this, regularly review AI-generated content and targeting for potential bias, and strive for inclusivity in your marketing.

Transparency: While you do not need to explain the intricate workings of every AI algorithm, be generally transparent when AI is significantly involved in interactions, such as with an AI chatbot. Some jurisdictions may require disclosure.

Authenticity and Human Oversight: AI-generated content should always be reviewed and refined by a human to ensure accuracy, alignment with your brand voice, and appropriateness. Do not let AI completely replace the human touch, especially in customer service and relationship building.

Job Displacement Concerns: While AI can automate tasks, focus on how it can augment human capabilities, freeing up your team for more strategic, creative, and interpersonal work, rather than solely as a means of replacing human roles.

Checklist: Key Ethical Questions for AI in Pickleball Marketing

Are we transparent with our customers about how their data is being used by AI?

Have we taken steps to minimize potential bias in our AI-driven targeting and content?

Is our AI-generated content accurate, truthful, and reflective of our brand values?

Do we have human oversight processes in place for AI-driven marketing activities?

Are we using AI to enhance customer experience, not just to cut costs?

How are we addressing any concerns about data security related to the AI tools we use?

Developing an AI-infused marketing strategy is an iterative process. Start with clear goals, strategically integrate AI into your customer's journey, budget wisely, and always prioritize ethical considerations. By doing so, you can harness the immense power of AI to not only grow your pickleball business but also to build stronger, more meaningful relationships with your customers. The following chapters will delve deeper into specific AI applications for core marketing functions, all guided by the strategic principles outlined here.

CHAPTER 4

AI for Prospecting and Lead Generation

For any pickleball business to achieve sustained growth, a consistent influx of new customers (B2C) or clients (B2B) is fundamental. Prospecting, which involves identifying potential customers, and lead generation, the process of attracting and converting these prospects into interested leads, are foundational to sales and expansion. Traditionally, these activities could be time-consuming and often involved less targeted approaches. Artificial Intelligence is transforming this domain, offering tools and techniques to make prospecting more focused, lead generation more efficient, and outreach efforts more personalized and effective.

This chapter will explore how AI can be a valuable asset in identifying high-potential prospects, crafting compelling outreach strategies, automating aspects of lead qualification, and tailoring these approaches for specific pickleball business scenarios, such as club membership drives, B2B software sales, or new product launches.

Identifying Potential Customers (B2C) and Clients (B2B) with AI Tools

The initial step in prospecting is determining who to target. AI can help you identify individuals or businesses that align with your ideal customer profile (developed in Chapter 2) with significantly greater precision than manual methods.

AI-powered sales intelligence platforms can sift through vast amounts of public data—including company websites, social media activity, news articles, and professional networks like LinkedIn—to identify businesses or individuals matching specific criteria. These tools can assist in finding B2B clients (e.g., pickleball clubs in a particular region that do not yet have an online booking system, if you offer such software) or even identifying B2C demographic segments demonstrating high interest in pickleball.

Even without specialized sales intelligence tools, LLMs like ChatGPT or Claude can serve as powerful research assistants. You can prompt them to search for specific types of businesses or to analyze information to identify potential leads.

Prompt Example (for ChatGPT/Claude, for B2B prospecting):

"Act as a market research assistant. I'm seeking potential B2B clients for my company, which sells custom-branded pickleball apparel to teams and clubs.

Identify 10 pickleball clubs or leagues in the state of [State Name] that meet the following criteria:

– Appear to have a large and active membership base OR regularly host tournaments or events (2023–2024)

– Show signs of recent activity (e.g., social media, updated website)

– **Do not** have an existing online store selling branded apparel (on-site or via third-party vendors)

For each organization, provide:

– Name of the club or league

– Website URL (if available)

– A brief note on what makes them a strong B2B lead (e.g., "hosts 5+ tournaments per year," "1,000+ members")"

For more complex or ongoing prospecting research, an AI agent like Manus can be exceptionally effective. You can define a detailed set of criteria for your ideal B2B client or B2C customer segment, and Manus can be tasked to systematically search online sources, compile lists of potential prospects, and even gather initial contact information or identify key decision-makers within organizations.

Example Task for Manus (B2B Prospecting for a Court Resurfacing Company):

"Research and compile a list of 20 pickleball facilities (clubs, community centers, private courts) within a 100-mile radius of [Your City, State] that were established in 2019 or earlier.

Identify those whose **online reviews or website photos** suggest their courts may be due for resurfacing. Look for signs such as cracks, faded lines, uneven surfaces, or safety concerns mentioned in reviews or visible in imagery.

Prioritize facilities that host regular tournaments or leagues. For each, provide:

1. Facility Name
2. Address
3. Website URL
4. Contact Person or Department (e.g., Facility Manager or Operations Head)

5. Notes from reviews or site content regarding court condition

Deliver the list in a **CSV file.**"

AI-Powered Outreach Strategies (Personalization at Scale)

Once prospects are identified, the next step is outreach. Generic, impersonal messages are rarely effective. AI excels at enabling personalization at scale, allowing you to tailor your communication to resonate with each prospect's specific situation, needs, or interests.

LLMs can draft outreach emails that go far beyond simple name token replacement. By providing the AI with information about the prospect (e.g., their business, recent achievements, expressed interests), it can generate highly relevant and engaging messages.

Prompt Example (for ChatGPT/Claude, after researching a specific pickleball club):

"Draft a personalized outreach email to [Club Manager Name], manager of [Pickleball Club Name]. I represent a company offering AI-powered court scheduling software that streamlines booking and reduces administrative workload for busy clubs.

Based on their website (which notes a manual booking system and a recent large tournament), the email should:

1. Acknowledge the recent event positively.
2. Introduce the software and its main benefit: optimizing court usage and simplifying scheduling.

3. Gently mention the challenges of manual systems, especially as demand grows.
4. Propose a 15-minute demo call to explore how the software can help their club specifically.

Keep the message **personalized, concise, and benefit-oriented**. If helpful, include a brief note of social proof (e.g., "similar clubs are already seeing results")."

For a more sophisticated approach, an AI agent like Manus can manage entire outreach sequences. This could involve sending an initial personalized email, scheduling follow-up emails if there is no response, and even varying the content of follow-ups based on any engagement with previous messages. Manus can integrate data from its prospecting research to ensure each message in the sequence is as relevant as possible.

Automating Lead Qualification with AI Chatbots and CRMs

Not every prospect who shows initial interest will be a suitable fit for your business. Lead qualification is the process of determining which leads are most likely to convert into paying customers. AI can automate significant parts of this process.

AI chatbots on your website or social media can engage visitors in conversation, ask pre-programmed qualifying questions (e.g., about their budget, needs, timeline, or, for a club, their playing level and availability), and categorize leads based on their responses. This ensures that your sales team (or you) spends time on the most promising leads.

Tool Spotlight: AI Chatbot Platforms

* **Strengths:** Platforms such as Tidio, Drift, Intercom, or simpler tools with AI capabilities provide 24/7 engagement,

instantly answer common questions, capture lead information, and perform initial qualification based on defined rules.

* **Example Script Outline for an AI Chatbot (Pickleball Facility Website):** 1. Greeting: "Welcome to [Facility Name]! How can I assist you today? (Options: Book a Court, Membership Information, Lessons, Other Inquiries)"

2. If "Membership Information": "Excellent! Are you new to pickleball, or an experienced player?"

3. Based on response: Provide relevant membership package information and ask, "Would you like to schedule a tour or speak with our membership coordinator?"

4. Capture contact details if they agree.

Many modern Customer Relationship Management (CRM) systems incorporate AI to score leads based on their demographics, engagement with your marketing materials (website visits, email opens, content downloads), and other factors. This helps prioritize which leads sales personnel should focus on.

An AI agent like Manus can take lead qualification further. For instance, if a chatbot identifies a moderately interested lead, Manus could trigger a nurturing sequence, sending them relevant content (e.g., blog posts, case studies, testimonials) over a period, and then re-evaluate their interest level based on their engagement with this content before passing them to a human salesperson.

Specific AI Strategies & Prompts for Pickleball Business Scenarios

Let's apply these concepts to common pickleball business goals:

Club Membership Drives:

Prospecting: Use Manus to monitor local community Facebook groups or neighborhood forums for phrases like "looking for pickleball lessons," "where to play pickleball in [City]," or "beginner pickleball." Task Manus to compile a list of individuals expressing interest

Outreach: Use ChatGPT to draft friendly, non-intrusive messages inviting these individuals to an open house or a free introductory clinic at your club. Manus could help manage sending these invites and tracking responses.

Qualification: An AI chatbot on your club's website can answer questions about membership tiers, clinic schedules, and facility amenities, qualifying leads based on their stated interests and playing level.

AI Workflow for this Membership Drive – 3 Prompts

Prompt Example (for Manus to Monitor groups and forums):

"Monitor public online forums and community platforms (such as Facebook Groups, Reddit, and Nextdoor) for posts within a 25-mile radius of [Your City, State] that include phrases like 'looking for pickleball lessons,' 'where to play pickleball in [City],' 'beginner pickleball,' or similar expressions of interest.

For each post that matches, extract and compile the following information into a Google Sheet or CSV:

– Name or username of the poster

– Date of the post

– Full text of the post

– Platform/source (e.g., Facebook, Reddit)

– Direct link to the post (if available)

Update the list weekly, and prioritize posts that show clear intent to join, learn, or play pickleball locally."

Prompt Example (for ChatGPT to draft friendly, non-intrusive messages):

"Draft a friendly, conversational message I can send to individuals who recently posted online about wanting to learn or play pickleball in [Your City]. The message should invite them to a free introductory clinic or open house at our club. Keep the tone warm and helpful, not salesy. Mention that spots are limited to encourage early response, and include a simple call-to-action (e.g., reply to RSVP or click a link to register)."

Prompt Example (for Manus to manage sending and tracking):

"Send the message below to individuals listed in our recent prospecting sheet who expressed interest in pickleball. Use their preferred contact method (email or messaging, if available). Track who opens the message, who responds, and who clicks the RSVP link. Update the outreach tracking sheet daily with status labels: 'Sent,' 'Opened,' 'Replied,' 'Clicked RSVP,' or 'No Response.'"

B2B App Sales to Clubs (e.g., a new pickleball tournament management app):

Prospecting: Task Manus to research and identify 50 pickleball clubs or tournament organizers across the country who frequently host events but appear to be using manual or outdated systems for registration and bracketing (based on their websites or event pages).

- **Prompt Example (for Manus):**

"Research and compile a list of 50 pickleball clubs, facilities, or tournament organizers across the U.S. that regularly host tournaments, leagues, or competitive events. Focus on organizations that appear to be using manual or outdated systems for registration, scoring, or bracket management—such as Google Forms, printable PDFs, spreadsheets, or handwritten brackets (as seen on their websites or event pages).

For each organization, provide the following in a CSV file:

1. Club or organization name
2. Website URL
3. Event type(s) hosted (e.g., leagues, tournaments)
4. Evidence of outdated systems (e.g., screenshots, wording, or links to forms)
5. Contact person or department (e.g., Tournament Director, Events Coordinator)
6. Location (City, State)

Prioritize organizations with recurring events or large participant numbers. Include a column for notes on any pain points or inefficiencies observed during your research."

Outreach Prompt (for ChatGPT/Claude, to be used by Manus in an email sequence): "Subject: Streamline Your

[Club Name/Tournament Series Name] Events with AI-Powered Management

Dear [Contact Person],

I came across your impressive [mention specific tournament or event] and noted the vibrant pickleball community you are fostering.

Managing successful tournaments involves many operational details. Our new AI-enhanced platform, [App Name], is designed specifically for pickleball event organizers like you to simplify registrations, automate bracket creation, manage player communication seamlessly, and provide real-time scoring updates – saving significant administrative time.

Would a brief 15-minute virtual demonstration be helpful to show you how [App Name] could make your next event smoother and more professional?

Sincerely, [Your Name]"

Qualification: Through an API Manus tracks email opens and clicks. If a lead clicks a link to a demo video, Manus flags them as a warmer lead and schedules a follow-up task for a human salesperson. **Prompt:** See Sending and Tracking Prompt above)

New Product Launches to Consumers (e.g., a new type of pickleball paddle):

Prospecting: Use AI analytics on your existing customer base to identify early adopters or those who have purchased similar premium products in the past. Use social listening AI to find influencers or avid players discussing needs your new paddle addresses.

Outreach: Craft targeted email campaigns (using AI for copy and segmentation) to these identified segments, announcing the new paddle and offering an exclusive early bird discount.

Qualification/Engagement: Use an AI-powered quiz on your website: "Find Your Perfect Paddle!" Based on their answers about playing style, experience, and preferences, the quiz recommends the new paddle if it is a fit, guiding them towards a purchase.

By strategically employing AI in your prospecting and lead generation efforts, you can build a more robust and predictable sales pipeline. AI tools, from LLMs for research and copy to sophisticated autonomous agents like Manus for workflow automation, empower you to find the right audience, reach out with compelling messages, and efficiently qualify leads, ultimately driving more conversions and growing your pickleball business. The key is to combine AI's analytical power with a genuine understanding of your prospects' needs and a human touch where it matters most.

CHAPTER 5

Crafting Compelling Content with AI

In today's digital environment, high-quality content is a critical asset for any business. For your pickleball enterprise, engaging and informative content serves as the cornerstone for attracting, educating, and retaining customers. Whether it is compelling website copy, insightful blog posts, shareable social media updates, or persuasive product descriptions, your content significantly shapes brand perception and influences customer actions. However, consistently producing a high volume of diverse and engaging material can be a substantial challenge, particularly for busy entrepreneurs and small marketing teams. Artificial Intelligence can serve as a powerful co-creator and amplifier in this process.

This chapter will explore the various ways AI can enhance your content creation workflow. We will cover AI applications for brainstorming and idea generation, writing different forms of text-based content, creating visually appealing graphics, exploring AI-assisted video creation, and, importantly, how to maintain your unique brand voice while leveraging these advanced tools. Throughout this chapter, we will highlight specific AI tools and provide actionable prompt examples tailored for pickleball businesses. An AI agent like Manus can also play a role in orchestrating and automating parts of the content lifecycle.

AI for Brainstorming and Idea Generation

Effective content originates from strong ideas. However, even creative individuals can encounter challenges in generating fresh perspectives. AI, particularly LLMs, can be a valuable resource for inspiration.

By providing an LLM with information about your target audience (from Chapter 2), your business objectives, and current pickleball trends (from Chapter 1), you can generate a wide range of content ideas.

Prompt Example (for ChatGPT/Claude):

"Act as a content strategist for a pickleball coaching business that helps intermediate players (3.0–4.0 level) improve their strategy and mental game. Our audience values actionable tips, deeper insight into gameplay decisions, and relatable stories of improvement.

Generate 20 original content ideas suitable for:

- Long-form blog posts (~1000 words)
- 30–60 second Instagram Reels (scriptable moments or visual tips)
- Thoughtful LinkedIn articles targeting serious recreational players
- A weekly newsletter tip section

For each idea, provide:

– A concise, catchy title

•

– A 1–2 sentence description of the content angle and which format it's best suited for Ensure ideas are **distinct**, balance both **strategic and mental game challenges**, and reflect our tone: clear, encouraging, and experience-based."

AI can also help identify trending topics or questions your audience is asking online, ensuring your content is relevant and timely. Manus could be tasked to monitor specific pickleball forums or Q&A sites and identify frequently asked questions, which can then inform your content calendar.

Writing Engaging Website Copy and Product Descriptions with AI

Your website often creates the first impression a potential customer has of your pickleball business, and your product descriptions function as digital sales representatives. AI can help ensure both are clear, compelling, and optimized for conversion.

LLMs can draft copy for your homepage, about us page, service pages, and FAQs. You provide the key information, target audience, and desired tone, and the AI generates a draft for your refinement.

Tool Spotlight: ChatGPT & Claude

* **Strengths:** Both are excellent for generating various forms of text. ChatGPT is often noted for its versatility and speed, while Claude is recognized for its nuanced writing, strong reasoning capabilities, and ability to handle longer, more complex prompts, making it particularly suitable for thoughtful or detailed copy.

* **Ideal Use Cases:** Drafting website pages, blog posts, email sequences, social media captions, and scripts.

Prompt Example (for ChatGPT/Claude - About Us Page):

"Draft the "About Us" page copy for a new e-commerce brand called **Pickleball Performance Pro**, which sells high-performance pickleball paddles, shoes, and accessories. Our customers are competitive and serious recreational players who care deeply about their game and their gear.

Our brand values are **innovation, quality**, and a deep understanding of what pickleball players truly need. The tone should be **authoritative yet passionate and inspiring**—we're not just selling gear; we're helping players perform at their best.

Please structure the copy with:

– A strong opening paragraph that draws in performance-minded players

– A clear statement of our mission or philosophy

– A short paragraph about our commitment to testing, curating, and sourcing only the best equipment

– A personal note that the founders are avid players themselves and built this store to meet a gap they saw in the market

Highlight what makes our store different, such as expert guidance, premium selection, and a shared passion for the sport. Aim for approximately 400 words."

For e-commerce businesses, persuasive product descriptions are vital. AI can take a list of product features and transform them into benefit-driven copy that appeals to your target customer.

Prompt Example (for ChatGPT/Claude - Pickleball Paddle Description):

"Write a compelling and detailed product description for a new carbon fiber pickleball paddle named the" ShadowStrike X1." The target audience is intermediate to advanced players seeking a balance of power and control.

Key Features: * Face Material: T700 Raw Carbon Fiber with Grit Surface * Core: 16mm Reactive Polymer Honeycomb * Weight: 7.9 - 8.2 oz (Mid-weight) * Grip: Cushioned, perforated, 4.25-inch circumference * Shape: Elongated with a slightly tapered head * Sweet Spot: Generous, enhanced by core technology

The description should: 1. Have a catchy headline. 2. Translate each feature into a clear player benefit (e.g., raw carbon for maximum spin, 16mm core for enhanced control and feel). 3. Emphasize the paddle's suitability for competitive play. 4. Use strong, evocative language. 5. Include a call to action (e.g., "Elevate your game with the ShadowStrike X1 today!"). Aim for 200-250 words."

Creating Eye-Catching Visuals with AI Design Tools

Visual content is crucial for capturing attention, especially on social media and websites. AI-powered design tools make it easier to create professional-looking graphics, even without extensive design experience.

Tool Spotlight: Canva AI

* **Strengths:** Features an extremely user-friendly interface with a vast library of templates. Its AI capabilities (Magic Media for text-to-image, Magic Write for generating text within designs, Magic Design for instant template suggestions based on your content) significantly speed up the creative process for social media posts, flyers, presentations, and more.

* **Ideal Use Cases:** Creating social media graphics, email headers, website banners, simple marketing videos, presentations, and flyers for pickleball events or promotions.

Prompt Example (for Canva Magic Media - Text-to-Image):

"Create a vibrant, energetic image for an Instagram post announcing a weekend pickleball clinic. Show a diverse group of adults (ages 30–60) smiling and playing pickleball outdoors on a sunny day. Include modern, colorful athletic wear appropriate for warm weather. The style should be photorealistic with a slightly artistic, dynamic feel. Emphasize motion, fun, and the spirit of learning—some players should be mid-action while others are observing or laughing nearby."

Other text-to-image generators like DALL-E, Midjourney, and Adobe Firefly offer more advanced capabilities for generating unique images from text prompts, suitable for custom illustrations or conceptual visuals.

Video Content Creation and Editing with AI

Video is one of the most engaging forms of content. While complex video production still requires significant skill, AI is simplifying aspects of video creation and editing.

LLMs can draft scripts for short promotional videos, instructional clips, or social media reels.

Prompt Example (for ChatGPT/Claude - Short Video Script):

"Write a script for a 45-second Instagram Reel demonstrating the proper technique for a pickleball dink shot. The target audience is beginner to early intermediate players. The script should include: 1. A catchy hook/intro (5 seconds). 2. Clear, concise instructions for stance, grip, swing, and ball contact (30 seconds, broken into 2-3 visual cues). 3. A brief outro with a call to action (e.g., "Follow for more pickleball tips!") (10 seconds). The language should be encouraging and easy to understand. Indicate where visuals or demonstrations would occur."

AI voiceover tools can generate natural-sounding voiceovers from text scripts. Some AI video editing tools offer features for automatic scene detection, highlight reel creation from longer footage, or even generating simple videos from text articles or blog posts (e.g., Pictory.ai, InVideo AI). AI avatars can create videos featuring AI-generated characters speaking your script, useful for presentations or informational content without requiring an on-camera presence.

Maintaining Your Brand Voice with AI: Tips and Best Practices

While AI is a powerful content creator, it is crucial that the output aligns with your unique brand voice and values. AI-generated content should typically serve as a starting point, not always the final product.

Develop a Clear Brand Voice Guide: If you have not already, document your brand's personality, tone (e.g., playful, professional, instructional, empathetic), preferred terminology, and elements to avoid. This guide can inform any AI tool or human writer.

Use Custom Instructions/Prompts: Many LLMs (like ChatGPT) allow users to set custom instructions that influence their responses. You can inform the AI about your brand voice, target audience, and desired output style in these instructions.

Provide Specific Context in Prompts: The more detailed your prompt, the better the AI can match your desired style. Include information about the target audience, the purpose of the content, the desired tone, and any key messages.

Iterate and Refine: Do not expect the first AI output to be perfect. Use it as a draft. Edit, rephrase, and add your unique insights and personality. Request revisions from the AI based on your feedback (e.g., "Make this sound more energetic," or "Rewrite this for a beginner audience").

Human Oversight is Essential: Always review AI-generated content for accuracy, factual correctness, tone, and brand alignment before publishing. AI can sometimes make errors or produce generic-sounding text if not guided properly.

Train Your AI (Where Possible): Some advanced AI platforms allow for fine-tuning on your own content, helping them learn your specific style more deeply over time.

Manus in the Content Lifecycle

An AI agent like Manus can play a significant role in orchestrating your content strategy:

* **Content Calendar Management:** Task Manus to help manage your content calendar, suggest topics based on

ongoing research (trends, competitor activity, community discussions), and even draft initial outlines or briefs for writers (human or AI).

* **Content Repurposing:** Manus could be tasked to take a long-form blog post and suggest ways to repurpose it into social media snippets, an email newsletter section, or a script for a short video, then even generate initial drafts for these repurposed formats.

* **Performance Monitoring:** Manus can be assigned to gather performance data for your published content (e.g., website traffic for blog posts, engagement rates for social media) and compile reports, helping you understand what resonates with your audience and refine your content strategy accordingly.

By thoughtfully integrating AI into your content creation workflow—from initial brainstorming to final publication and performance analysis—your pickleball business can produce a higher volume of more targeted, engaging, and effective content, all while saving valuable time and resources. Remember, AI is your collaborator, not a replacement for your unique brand perspective and human creativity.

CHAPTER 6

Supercharging Your Email Marketing with AI

Email marketing remains one of the most effective and direct methods for reaching your audience, nurturing leads, and driving conversions for your pickleball business. Unlike social media algorithms that can limit your reach, email lands directly in your subscribers' inboxes, offering a personal channel for communication. However, crafting engaging emails, managing lists, personalizing messages at scale, and analyzing performance can be notably time-consuming. Artificial Intelligence offers a suite of tools to enhance every aspect of your email marketing, making it more efficient, personalized, and impactful.

This chapter will guide you through leveraging AI to build and segment your email lists, write high-converting email copy for various campaign types, personalize your messaging at scale, utilize AI for A/B testing, and automate complex email workflows. We will provide specific tool insights and actionable prompt examples, demonstrating how an AI agent like Manus can orchestrate and enhance these processes for your pickleball business.

Building and Segmenting Your Email List with AI

A healthy, engaged email list is the foundation of successful email marketing. AI can assist in both growing your list and segmenting it effectively for targeted communication.

AI can help brainstorm and even draft lead magnets, such as a short e-book on "Five Common Pickleball Mistakes and How to Fix Them," a checklist for "Preparing for Your First Pickleball Tournament," or a discount code for first-time purchasers at your online store. Many website tools now use AI to optimize the timing and display of email sign-up forms or pop-ups based on visitor behavior, increasing conversion rates.

Once you have subscribers, sending the same message to everyone is rarely optimal. AI can analyze your subscriber data (demographics, purchase history, email engagement, website activity, stated interests from sign-up forms) to identify meaningful segments. This allows you to send more relevant content to each group. For example, an online pickleball store could use AI to segment its list into "Beginner Players," "Intermediate Players," "Advanced Players," "Pickleball Coaches," and "Gift Buyers." Each segment would then receive tailored newsletters and promotions.

An AI agent like Manus can be tasked to automate parts of list management. For example, based on predefined rules or engagement data (like consistent non-opens), Manus could help identify inactive subscribers for a re-engagement campaign or for list cleaning, ensuring your list remains healthy and your deliverability rates stay high. It could also automatically segment new subscribers based on information they provide during sign-up or their initial interactions with your website.

Writing High-Converting Email Copy with AI

Crafting email copy that grabs attention, conveys your message clearly, and persuades readers to take action is an art. AI, particularly LLMs, can be an invaluable assistant in this process, helping you write everything from subject lines to full email bodies.

AI can generate multiple subject line variations, often incorporating elements of curiosity, urgency, or benefit.

Prompt Example (for ChatGPT/Claude):

"Generate 10 compelling email subject lines for a weekly newsletter from a pickleball club. The newsletter typically includes club news, upcoming event announcements, a pickleball tip of the week, and member spotlights. Vary the tone – include options that are informative, exciting, and community-focused. Ensure some are short and punchy, and others are slightly more descriptive."

Whether it is a welcome email, a promotional offer, a newsletter, or an abandoned cart reminder, AI can draft the entire email.

Tool Spotlight: ChatGPT & Claude

* **Strengths:** Excellent for drafting various email types. Provide them with the goal of the email, the target segment, key information to include, and the desired tone.

Prompt Example (for Claude - Welcome Email Series):

"Act as an email marketing specialist for "PicklePro Gear," an online store selling premium pickleball equipment. Draft a three-part welcome email series for new newsletter subscribers.

The goal is to introduce the brand, build trust, and encourage their first purchase.

Email 1 (Sent immediately after sign-up):

* Subject: Welcome to the PicklePro Gear Family! Your 10% Off Code Inside! * Content: Provide a warm welcome, briefly introduce PicklePro Gear (emphasizing passion for quality and a curated selection), offer a 10% discount code for their first order, and set expectations for future newsletters (expert tips, new arrivals, exclusive deals).

Email 2 (Sent 2 days later): * Subject: Find Your Perfect Paddle with PicklePro Gear * Content: Highlight a key product category (e.g., paddles), offer advice on choosing the right paddle (linking to a relevant blog post or guide), and showcase two to three popular paddles with brief descriptions and benefits.

Email 3 (Sent 4 days later): * Subject: Don't Miss Out on These Pickleball Essentials! * Content: Showcase other essential gear (balls, bags, apparel), perhaps feature a customer testimonial, and gently remind them of their 10% discount code: If you haven't ordered yet, there's still time to use your 10% off code!

Maintain a friendly, helpful, and enthusiastic tone throughout the series. Use short paragraphs, optional bullet points for product highlights, and end each email with a clear CTA."

Prompt Example (for ChatGPT - Promotional Email):

"Draft a promotional email for a 48-hour flash sale on all pickleball shoes at "CourtSideKicks.com." The sale offers a 25% discount. The email should: **1.** Have an urgent and exciting subject line. **2** Clearly state the offer (25% off all shoes) and the duration (48 hours only). **3.** Highlight the benefits of upgrading pickleball shoes (e.g., better grip, support, injury prevention). **4.** Showcase two to three popular shoe models with their original and sale prices. **5.** Include a clear call-to-action button: "Shop the Sale Now!" **6.** Create a sense of urgency to encourage immediate purchase. The tone should be energetic and persuasive."

Personalizing Email Campaigns at Scale with AI

Personalization extends beyond simply using the subscriber's first name. True personalization involves tailoring the content, offers, and timing of your emails to individual subscriber preferences and behaviors. AI makes this level of personalization achievable at scale.

Many email marketing platforms now offer AI features that can dynamically change parts of an email (e.g., product recommendations, images, calls to action) based on individual subscriber data or segments. For e-commerce businesses, AI can analyze a subscriber's past purchase history and browsing behavior to recommend products they are most likely to be interested in, directly within an email. AI can also trigger automated emails based on specific subscriber actions (or inactions), such as visiting a particular page on your website, abandoning a shopping cart, or not opening emails for a certain period.

AI for A/B Testing Subject Lines and Content

A/B testing (or split testing) is crucial for optimizing your email performance. It involves sending two or more variations of an email to different subsets of your list to see which performs better (e.g., in terms of open rates or click-through rates). AI can assist in generating variations and analyzing results.

AI can quickly create multiple versions of subject lines, calls to action, or even entire email bodies for testing.

Prompt Example (for ChatGPT/Claude):

"Generate five alternative subject lines for A/B testing, each using a different psychological angle — one focused on **urgency**, one on **benefits**, one on **curiosity**, one on **scarcity**, and one on **social proof**. Keep the tone fun and relevant to recreational pickleball players."

This makes it easy to see which angle performs best, and it adds a fifth proven angle (social proof).

Some advanced AI tools claim to predict which variations are likely to perform best even before you send them, potentially saving time and resources. While most email platforms provide A/B test analytics, AI can help you dig deeper into the results, perhaps identifying why one version outperformed another beyond just the numbers.

Automating Email Workflows with AI (and Manus)

Email workflows (also known as drip campaigns or automated sequences) are series of emails sent automatically based on specific triggers or timelines. AI enhances these workflows by enabling more complex logic and personalization.

Common automated workflows include welcome series, abandoned cart reminders, re-engagement campaigns for inactive subscribers, and post-purchase follow-ups (e.g., asking for a review, offering tips on using the purchased product), and birthday/anniversary emails.

An AI agent like Manus can take email automation to the next level by orchestrating workflows that span multiple channels and involve more sophisticated decision-making. For example, if a prospective member for a pickleball club abandons the online sign-up form, this could trigger Manus to: send an automated follow-up email with a special offer; if no response, schedule a task for a club representative to make a personal phone call; if the prospect mentioned interest in beginner lessons during an earlier website chat (data captured by an AI chatbot), Manus could ensure the follow-up email specifically highlights upcoming beginner clinics; if the prospect then signs up, Manus triggers the standard welcome email series and adds them to the "New Members" segment for future targeted communications.

By integrating AI into your email marketing, you transform it from a potentially laborious task into a highly efficient, personalized, and data-driven engine for growth. From building your list and crafting compelling copy to automating sophisticated workflows with tools like Manus, AI empowers your pickleball business to connect more effectively with your audience, nurture relationships, and drive the actions that matter most. Remember to always combine AI's power with authentic brand messaging and a genuine desire to provide value to your subscribers.

CHAPTER 7

Leveraging AI for Social Media Engagement

Social media is an undeniable force in modern marketing, and for the vibrant, community-driven world of pickleball, it represents an essential arena for engagement, brand building, and customer acquisition. Platforms such as Instagram, Facebook, TikTok, X (formerly Twitter), and even LinkedIn (for B2B interactions) offer significant opportunities to connect with players, fans, clubs, and businesses. However, managing a successful social media presence—which includes creating consistent content, engaging with followers, running effective ad campaigns, and analyzing performance—can be a demanding undertaking. Artificial Intelligence provides a wealth of tools to streamline these efforts, making your social media marketing more strategic, efficient, and impactful.

This chapter will explore how AI can help you develop a data-driven social media strategy, optimize content creation and scheduling, engage your community more effectively, run powerful AI-targeted ad campaigns, and measure your social media return on investment (ROI). We will include practical examples and prompts tailored for pickleball businesses, and discuss how an AI agent like Manus can automate and enhance your social media operations.

Developing a Data-Driven Social Media Strategy with AI

A successful social media presence begins with a clear strategy. AI can help you move beyond guesswork by providing data-driven insights to inform your approach.

AI can analyze social media data to further refine your understanding of where your target pickleball audience spends their time online, what kind of content they engage with, and who the key influencers are in their communities, building upon the customer personas developed in Chapter 2. AI tools can also monitor your competitors' social media activities, identifying their most successful content, engagement strategies, and follower growth patterns. This can reveal opportunities and inform your own content pillars.

Prompt Example (for ChatGPT/Claude, after providing links to competitor social media profiles):

"Analyze the Instagram presence of these three competitor pickleball apparel brands: [Profile URL 1], [Profile URL 2], [Profile URL 3]. Identify: **1.** Common content themes or pillars they focus on. **2.** Types of posts that generate the highest engagement (likes, comments, shares). **3.** Their posting frequency and typical content formats (e.g., Reels, carousels, single images). **4.** Their overall brand voice and visual style on Instagram. **5.** Any apparent gaps or opportunities in their content strategy that my brand could leverage. Present your findings in a comparative table followed by a short strategic summary with actionable recommendations."

Additionally, AI tools can track trending hashtags, topics, and content formats within the pickleball niche, helping you create timely and relevant posts.

AI for Content Creation, Scheduling, and Optimization

As discussed in Chapter 5, AI is a significant aid for content creation, and this extends powerfully to social media.

LLMs can quickly draft engaging captions, suggest relevant hashtags, and brainstorm post ideas tailored to different platforms and pickleball scenarios.

Prompt Example (for ChatGPT/Claude - Instagram Post for a Club):

"Our pickleball club, "The Dinking Dragons," just hosted a successful charity tournament. Draft three engaging Instagram caption options for a photo carousel showcasing highlights from the event (smiling players, action shots, check presentation). The tone should be celebratory, community-focused, and appreciative of participants and sponsors. Include relevant hashtags like #PickleballForGood #[ClubName] #[CityName]Pickleball #CharityEvent #PickleballCommunity. Each caption should be 1–2 short paragraphs max and include a natural call to action (e.g., 'Tag a teammate,' 'Thanks to everyone who made this happen,' etc.)."

Tools like Canva AI are invaluable for quickly designing eye-catching graphics, quote cards, event announcements, and even short video clips for social media. Many social media management platforms (e.g., Buffer, Hootsuite, Sprout Social) now incorporate AI to suggest optimal posting times by analyzing when your specific audience is most active, curate relevant third-party content to share, and recommend relevant hashtags to improve discoverability.

AI can also assist in content repurposing. For instance, Manus could be tasked to take a newly published blog post from your website, generate three to five key takeaway snippets suitable for X (formerly Twitter), draft a more visual post idea for Instagram, and then add these to a draft content schedule for your review.

Engaging Your Community with AI-Powered Tools

Social media is inherently interactive. Engagement—responding to comments, answering questions, and participating in conversations—is crucial for building a loyal community.

AI tools can track mentions of your brand, relevant keywords, or hashtags across social platforms, alerting you to conversations you should join or comments that need a response. For common questions or comments, LLMs can help draft initial responses, saving you time. However, always review and personalize these before posting to maintain an authentic voice.

Prompt Example (for ChatGPT/Claude - Responding to a common question):

"A potential customer asked on our Facebook page: "What are your pickleball club membership rates for families?" Draft a friendly and informative response that: 1. Thanks them for their interest. 2. Briefly explains that we offer family packages. 3. Directs them to the membership page on our website for detailed pricing OR suggests they send a private message with their specific family size for a custom quote. 4. Invites further questions."

An AI agent like Manus can monitor your social media channels for mentions and comments. It could be programmed to automatically respond to very simple, common FAQs (e.g., "What are your opening hours?") using a predefined knowledge base. For more complex queries, negative sentiment, or urgent customer service issues, Manus could flag them and notify a human team member for a personalized response. This ensures timely engagement while filtering out routine interactions.

Running Effective AI-Targeted Social Media Ad Campaigns

Paid social media advertising (e.g., Facebook/Instagram Ads, LinkedIn Ads for B2B) is a powerful way to reach a highly targeted audience. AI is at the core of these platforms' ad targeting capabilities and can also assist in creating ad creatives.

Social media ad platforms use sophisticated AI algorithms to help you define and reach your ideal audience based on demographics, interests, behaviors, and lookalike audiences (users similar to your existing customers). You can use AI to generate variations of ad copy (headlines, body text, calls to action) and even visuals for your campaigns, allowing for rapid A/B testing to see what resonates best.

Prompt Example (for ChatGPT/Claude - Facebook Ad Copy):

"Draft Facebook ad copy for a campaign promoting a new line of eco-friendly pickleball paddles from" GreenSwing Paddles." The target audience is environmentally conscious pickleball players aged 25-55.

Ad Creative Elements Needed:

* **Headline (maximum 40 characters):** Provide 3 variations.

* **Primary Text (maximum 125 characters):** Provide 2 variations (highlighting eco-friendly materials and performance).

* **Call to Action Button:** Suggest appropriate options (e.g., Shop Now, Learn More).

The ad should evoke a sense of sustainability, quality, and love for the game. Mention that a portion of profits goes to environmental charities."

Ad platforms also use AI to automatically optimize your campaigns in real-time, allocating budget to the best-performing ads and audiences to maximize your ROI.

Measuring Social Media ROI with AI Analytics

Understanding the return on investment of your social media efforts is crucial. AI-powered analytics can provide deeper insights beyond basic metrics like likes and followers.

AI can assist with sentiment analysis to gauge the overall feeling towards your brand, help identify key engagement drivers (which content types or topics drive meaningful interaction), and contribute to attribution modeling to understand social media's role in conversions. Instead of manually logging into multiple platforms, an AI agent like Manus can compile a regular social media performance report, pulling key metrics from all your active platforms into a single summary, highlighting trends and areas for improvement.

By strategically integrating AI into your social media marketing, your pickleball business can create a more engaging, efficient, and results-driven presence. From developing a smarter strategy and creating compelling content with tools like LLMs and Canva AI, to automating monitoring and reporting with an agent like Manus, AI empowers you to connect authentically with the passionate pickleball community and turn social engagement into tangible business growth.

CHAPTER 8

Exploring AI in Direct Mail and Local Marketing

While digital marketing channels like email and social media often dominate marketing discussions, traditional methods such as direct mail and geographically focused local marketing still hold significant value for certain pickleball businesses. This is particularly true for those with a physical presence or a defined local service area, such as a new pickleball club, a local retail store specializing in pickleball gear, or a coaching service targeting a specific community. Artificial Intelligence can bring a new level of sophistication and efficiency to these more traditional marketing avenues.

This chapter will explore how AI can enhance your direct mail campaigns through hyper-personalization and help optimize your local marketing efforts, particularly in local Search Engine Optimization (SEO) and online review management. We will provide practical examples and prompts, and discuss how an AI agent like Manus can assist in managing and automating aspects of these local marketing initiatives for your pickleball business.

AI for Hyper-Personalized Direct Mail Campaigns

Direct mail, including postcards, flyers, or letters, can be an effective way to reach a targeted local audience, especially when digital fatigue is a factor. The key to making direct mail effective in the modern era is personalization. Generic mailers are often discarded, whereas personalized pieces that speak directly to the recipient's needs or interests have a much higher chance of engagement. AI can help achieve this hyper-personalization.

Similar to email list segmentation (Chapter 6), AI can analyze your customer data or purchased mailing lists (where ethically and legally sourced) to identify segments within a geographic area that are most likely to respond to a direct mail offer. For example, a pickleball club could segment local households by demographics (e.g., families with children, active retirees) or proximity to the club.

Local clubs can also avail themselves of geo-targeted intent-based audiences. Data brokers, such as RightOnTimeData.com, with access to lists of consumers who have searched for pickleball related gear in the past week can segment the list based on zip code and provide you with a hyper-targeted list of physical addresses of potential members within 30 mile radius. These lists can also provide data on demographics including household income, children, whether married, net worth and so much more.

LLMs can help you craft personalized copy for your mailers. By providing the AI with information about a specific segment or even (with appropriate data and consent) individual households, you can generate tailored messages and offers.

Prompt Example (for ChatGPT/Claude - Personalized Postcard for a New Club):

"Act as a marketing copywriter for" [Your City] Pickleball Hub, a new indoor pickleball facility opening next month. We are sending a direct mail postcard to households within a 3-mile radius.

For a segment identified as "Families with children aged 8-16," draft postcard copy (approximately 75 words for one side) that: 1. Announces the grand opening. 2. Highlights family-friendly features (e.g., junior programs, family membership discounts, safe environment). 3. Includes a special offer for this segment (e.g., "First junior clinic free with family sign-up!"). 4. Provides a clear call to action (e.g., "Visit our website at [URL] or scan the QR code to learn more!"). The tone should be welcoming, energetic, and family-oriented."

You could then create variations for other segments, such as "Young Professionals" (highlighting evening leagues and social events) or "Seniors" (emphasizing low-impact fitness and morning socials).

While direct AI design for print is still an emerging field, AI can analyze effective direct mail designs and provide recommendations on layout, imagery, and calls to action known to perform well for specific demographics. Some modern direct mail services are also beginning to incorporate AI to assist with targeting and offer variable data printing at scale, making hyper-personalization more feasible.

Optimizing Local SEO with AI for Pickleball Facilities and Stores

For any pickleball business with a physical location (clubs, stores, dedicated courts) or a defined service area (local coaches, mobile retail units), Local Search Engine Optimization (Local SEO) is critical. This involves ensuring your business appears prominently in local search results (e.g., "pickleball courts near me" or "pickleball store in [Your City]"). AI offers several ways to boost your Local SEO.

AI tools can help identify the specific keywords and phrases that people in your local area are using to search for businesses like yours. This extends beyond general pickleball terms to include location-specific queries.

Your Google Business Profile (GBP) is a cornerstone of Local SEO. AI can help you write compelling business descriptions, generate GBP posts, and answer GBP questions.

Prompt Example (for ChatGPT/Claude - GBP Description):

"Draft a Google Business Profile description (maximum 750 characters) for "Court Masters Pickleball Club" located in Springfield, IL. We offer eight indoor courts, a pro shop, lessons for all levels, leagues, and social events. Keywords to include are: "pickleball Springfield IL," "indoor pickleball courts," "pickleball lessons," and "pickleball leagues." The description should be inviting and highlight our key features."

AI can also help you develop website content (e.g., service pages, blog posts) that is specifically targeted to your local area by incorporating local landmarks, neighborhood names, and relevant local information naturally into your content.

Prompt Example (for ChatGPT/Claude - Localized Blog Post Idea):

"Generate a blog post title and outline for "Springfield Pickleball Pro Shop" targeting local customers. The post should focus on "Choosing the Right Pickleball Paddle for Illinois Weather Conditions (Considering Indoor and Outdoor Play)." Incorporate local relevance by mentioning common local playing environments or seasonal considerations in Springfield."

Keeping your business information (Name, Address, Phone number - NAP) consistent across all online directories is crucial for Local SEO. An AI agent like Manus could be tasked to periodically audit your key local listings for accuracy and consistency. If discrepancies are found, Manus could flag them for correction or, with appropriate integrations, even attempt to update them automatically.

AI for Managing Online Reviews and Local Listings

Online reviews on platforms like Google, Yelp, and Facebook are incredibly influential for local businesses. Managing your online reputation effectively is key, and AI can assist.

AI tools can monitor various review sites and alert you to new reviews, especially negative ones that require a prompt response. Responding to reviews (both positive and negative) demonstrates that you value customer feedback. AI can help draft professional and empathetic responses, saving you time while ensuring consistency.

Prompt Example (for ChatGPT/Claude - Responding to a Positive Review):

"A customer left the following 5-star review for our pickleball store, "The Paddle Rack": "Great selection and super helpful staff! Found the perfect paddle thanks to Sarah." Draft a warm and appreciative response to this review. Mention Sarah if appropriate and encourage the customer to visit again. Keep the tone professional and public-friendly, suitable for a Google Business review."

Prompt Example (for ChatGPT/Claude - Responding to a Negative Review):

"A customer left a 2-star review for our club: "Courts were too crowded, and the booking system was confusing." Draft a polite and constructive response that: 1. Thanks the customer for their feedback. 2. Acknowledges their concerns about crowding and the booking system. 3. Briefly mentions any steps being taken to address these issues (e.g., "We are working on optimizing court schedules and have a tutorial for the booking system on our website"). 4. Invites the customer to contact management directly to discuss their experience further. Avoid being defensive; aim for a resolution-oriented tone."

AI can also be used for sentiment analysis of reviews to identify common themes, strengths, and areas for improvement in your pickleball business. An AI agent like Manus could monitor review platforms. When a new review appears, it could perform an initial sentiment analysis. For positive reviews, it might draft a suggested thank-you response for your approval. For negative reviews, it could immediately flag it for your attention and provide a link, ensuring a swift response. It could also track if responses have been posted and remind you if any are outstanding.

While digital-first strategies are often emphasized, neglecting local marketing and targeted direct mail can mean missing out on a significant segment of your potential pickleball clientele.

By thoughtfully applying AI to personalize your outreach, optimize your local online presence, and manage your reputation, you can make these traditional channels more effective and efficient than ever before. This ensures that your pickleball business is visible and appealing to the customers right in your neighborhood, ready to engage with your offerings.

CHAPTER 9

AI for Customer Service and Retention in Pickleball

Acquiring new customers is essential for growth, but retaining existing ones is equally, if not more, crucial for the long-term success and profitability of your pickleball business. Happy, loyal customers not only provide repeat business but also become valuable advocates, spreading positive word-of-mouth. Excellent customer service is a cornerstone of retention, and Artificial Intelligence offers innovative ways to enhance service quality, personalize interactions, and proactively address customer needs, thereby fostering loyalty.

This chapter will explore how AI can transform your customer service operations and boost retention strategies. We will cover AI chatbots for instant support, personalizing customer communication, using AI for feedback collection and sentiment analysis to improve your offerings, and leveraging AI insights to build effective loyalty programs. As always, we will provide practical examples, prompts, and show how an AI agent like Manus can automate and enhance these customer-centric processes.

AI Chatbots for Instant Customer Support

In today's fast-paced world, customers expect quick answers to their queries. AI-powered chatbots can provide instant,

24/7 support on your website or social media channels, handling a large volume of common inquiries without human intervention.

Chatbots can be trained with a knowledge base of frequently asked questions about your pickleball club (e.g., opening hours, membership fees, court booking procedures), products (e.g., paddle specifications, return policies), or services (e.g., coaching session availability, tournament rules). They can guide users through processes like signing up for a newsletter, finding specific information on your website, or initiating a return for an e-commerce purchase. If a chatbot cannot resolve an issue, it can gather preliminary information from the customer and then seamlessly hand over the conversation to a human agent, providing them with context to expedite resolution.

Tool Spotlight: Various AI Chatbot Platforms

* **Strengths:** Platforms such as Tidio, Intercom, Drift, or custom solutions using LLM APIs provide immediate responses, reduce wait times, free up human staff for more complex issues, and are available 24/7.

* **Example: Designing a Knowledge Base Snippet for a Pickleball Tournament Chatbot:** * **Question:** "How do I register for the upcoming Summer Slam tournament?"

* **Chatbot Answer (to be programmed):** "You can register for our Summer Slam tournament directly on our website! Just head to [Tournament Page URL] and click the 'Register Now' button. You will need to provide your name, email, playing level (e.g., 3.0, 3.5, 4.0+), and partner's name if playing doubles. Registration closes on [Date]. Let me know if you have any other questions!"

The effectiveness of a chatbot depends on its knowledge base. An AI agent like Manus could be tasked to help maintain and update this knowledge base. For example, if your club changes its membership pricing or adds a new clinic, Manus could be instructed to update the relevant FAQ entries in the chatbot's backend or flag them for manual review and update, ensuring the chatbot always provides accurate information.

Personalizing Customer Communication with AI

Beyond initial support, AI can help you personalize ongoing communication with your existing customers, making them feel valued and understood.

For an e-commerce store selling pickleball gear, AI can trigger personalized follow-up emails after a purchase. This could include tips on how to care for their new paddle, links to relevant instructional videos, or an invitation to review the product. AI can also track customer milestones (e.g., membership anniversary at a club, birthday, achieving a certain spending threshold) and trigger automated congratulatory messages or special offers.

Prompt Example (for ChatGPT/Claude - Membership Anniversary Email):

"Draft a warm and appreciative email to a pickleball club member, [Member Name], on their one-year membership anniversary with "The Pickleball Palace." The email should: 1. Congratulate them on their anniversary. 2. Thank them for being a valued member of the community. 3. Briefly highlight a positive aspect of the club (e.g., "We hope you have enjoyed the friendly games and improved your skills over the past year!"). 4. Offer a small token of appreciation (e.g., "As a thank you, enjoy a free guest pass on your next visit or 10%

off in our pro shop this month!"). 5. Reinforce the club's commitment to providing a great pickleball experience. The tone should be personal and celebratory."

Based on a customer's past interactions (e.g., clinics attended, articles read on your blog, products purchased), AI can suggest relevant new content, events, or products, keeping them engaged with your brand.

AI for Feedback Collection and Sentiment Analysis

Understanding what your customers think about your products, services, and overall experience is vital for continuous improvement and retention. AI can streamline feedback collection and provide deep insights from the data gathered.

As discussed in Chapter 2, AI can help create and distribute targeted surveys to gather feedback at various touchpoints (e.g., after a coaching session, following a tournament, post-purchase). AI excels at analyzing unstructured text data from open-ended survey responses, online reviews, social media comments, and customer service interactions to determine the overall sentiment (positive, negative, neutral) and identify recurring themes, praise points, and areas of frustration.

Prompt Example (for ChatGPT/Claude, after providing customer service chat logs):

"Analyze the provided customer service chat logs from the past week for our pickleball equipment website. Identify: 1. The top three most common issues or questions raised by customers. **2.** The overall sentiment of customer interactions (mostly positive, mixed, mostly negative). **3.** Any recurring compliments about our service or products. 4. Any recurring complaints or suggestions for improving the customer experience. **5.** Provide two to three direct (anonymized) quotes that illustrate key findings. Organize the output into a structured summary with bullet points and short subheadings for each section."

An AI agent like Manus can be tasked to continuously monitor various channels (review sites, social media, dedicated feedback emails) for customer sentiment and feedback. It can then compile regular reports summarizing key trends, highlighting areas needing improvement, and even identifying customers who express strong positive sentiment (potential advocates) or strong negative sentiment (requiring proactive outreach for resolution).

Building Loyalty Programs with AI Insights

Loyalty programs are a proven way to encourage repeat business and reward your best customers. AI can make your loyalty programs smarter and more effective.

AI can analyze customer data to identify your most valuable customers (e.g., based on purchase frequency, total spend, engagement level) who should be prime candidates for loyalty rewards. Instead of a one-size-fits-all loyalty program, AI can help you tailor rewards and offers based on individual customer preferences and behavior. For example, a customer who frequently buys pickleball apparel might receive early access to new clothing lines, while a customer who regularly attends advanced clinics might get a discount on a pro workshop.

Some advanced AI models can analyze customer behavior to predict which customers are at risk of churning (i.e., ceasing to do business with you). This allows you to proactively reach out with special offers, address their concerns, or remind them of the value you provide, in an effort to retain them. An AI agent like Manus could potentially monitor customer engagement metrics (e.g., last visit to the club, last purchase, email open rates) and flag accounts that show signs of disengagement according to predefined rules, triggering a task for your team to initiate a personalized retention effort.

Excellent customer service and robust retention strategies are critical for the sustained success of your pickleball business. By leveraging AI for instant support, personalized communication, insightful feedback analysis, and smarter loyalty programs, you can create a customer experience that not only satisfies but delights, turning casual players or shoppers into lifelong fans and advocates. With tools like Manus helping to automate and orchestrate these efforts, you can build stronger customer relationships and a more resilient business in the competitive pickleball market.

CHAPTER 10

Measuring Success: AI for Marketing Analytics and Reporting

In the dynamic world of marketing, launching campaigns and creating content is only part of the process. To truly understand what is working, what is not, and how to optimize your efforts for better results, you need to measure your performance. Marketing analytics can often feel overwhelming, with a deluge of data points and metrics to track. Artificial Intelligence is transforming this landscape by not only making data collection more efficient but also by uncovering deeper insights, automating reporting, and enabling more data-driven decision-making for your pickleball business.

This chapter will focus on how AI can help you measure the success of your marketing initiatives. We will discuss key AI marketing metrics to track (moving beyond simple vanity metrics), explore AI tools for data visualization and reporting that make complex data understandable, and emphasize how to use these AI-powered insights to make informed decisions and continuously optimize your campaigns. We will also highlight how an AI agent like Manus can play a crucial role in automating the collection and compilation of these vital analytics.

Key AI Marketing Metrics to Track (Beyond Vanity Metrics)

While likes, followers, and page views (often called vanity metrics) can provide a surface-level indication of activity, they do not always tell the full story of your marketing effectiveness or its impact on your pickleball business's bottom line. AI can help you track and analyze more meaningful metrics that reflect true engagement, conversion, and return on investment (ROI).

Key metrics include Engagement Rate (analyzing the quality of interaction beyond likes), Conversion Rate (tracking actions like newsletter sign-ups or sales), Customer Acquisition Cost (CAC), Customer Lifetime Value (CLV), Return on Ad Spend (ROAS), Website Traffic Sources and Behavior, Lead-to-Customer Conversion Rate, and Sentiment Analysis.

AI Tools for Data Visualization and Reporting

Raw data can be difficult to interpret. AI-powered tools are increasingly adept at transforming complex datasets into easily understandable visual reports, dashboards, and summaries.

Many platforms you already use likely have AI enhancements. Google Analytics 4 (GA4) incorporates AI to provide predictive insights and anomaly detection. Social media platform analytics (Facebook/Instagram Insights, X Analytics, etc.) use AI to present data on audience and content performance. Email marketing platforms often use AI to surface key trends in open rates and click-through rates.

Dedicated AI analytics and Business Intelligence (BI) platforms like Tableau, Microsoft Power BI, and Google Looker Studio can connect to multiple data sources. They use AI to automate report generation, allow natural language queries (e.g., "What was our best-selling pickleball paddle category last month?"), detect anomalies in data, and provide predictive analytics.

Tool Spotlight: Google Looker Studio

* **Strengths:** Highly customizable for creating interactive dashboards that pull data from various Google services and other connectors. AI integration can enhance its ability to surface insights automatically.

* **Ideal Use Cases for Pickleball Businesses:** Creating a central marketing dashboard that shows website traffic, social media engagement, email campaign performance, and sales data (if integrated) all in one place.

Manus for Automated Multi-Source Reporting

One of the significant challenges in marketing analytics is consolidating data from disparate sources. An AI agent like Manus can be a powerful ally here.

Example Task for Manus (Weekly Marketing Performance Snapshot):

"`Every Monday morning, we will compile a Weekly Marketing Performance Snapshot for" The Pickleball Pro Shop." Gather the following data:

From Google Analytics 4: Total website users, new users, top three traffic sources, overall website conversion rate for newsletter sign-ups, and number of online paddle sales.

From Instagram Insights: Follower growth, average engagement rate per post, and reach of top three posts from the past week.

From Mailchimp (Email Marketing Platform): Open rate and click-through rate for the latest newsletter, and number of new subscribers.

From Shopify (E-commerce Platform): Total online sales revenue and average order value.

Present this data in a structured report with brief bullet-point summaries highlighting key changes or trends compared to the previous week. Identify any significant anomalies (e.g., a sudden 50% drop in website traffic from organic search).`"

This automated report saves significant time and ensures consistent monitoring of key metrics across all your important channels.

Making Data-Driven Decisions to Optimize Campaigns (Using Data to Improve Campaigns Over Time)

The ultimate goal of marketing analytics is not just to report on what happened, but to inform what you do next. AI-powered insights should fuel an iterative cycle of optimization: Measure, Analyze, Hypothesize, Test, and Learn & Iterate.

AI can help you identify correlations and patterns that might not be obvious to a human analyst, speeding up this optimization loop. For example, AI might find that customers who engage with your pickleball club's Instagram Reels showing member testimonials are 30% more likely to inquire about membership than those who only see static image posts.

By embracing AI for marketing analytics and reporting, your pickleball business can move from reactive observations to proactive, data-driven strategies. Understanding your performance through the lens of meaningful metrics, visualizing data for clarity, and using those insights to continuously refine your approach will ensure your marketing efforts are not just busy work, but are genuinely contributing to your growth and success in the competitive pickleball market. With tools like Manus automating the often-tedious data aggregation, you can focus more of your time on strategic interpretation and action.

CHAPTER 11

The Pickleball AI Toolkit Revisited (Expanded & Advanced)

Throughout this book, we have explored a multitude of ways Artificial Intelligence can revolutionize how you market and grow your pickleball business. We have touched upon various tools and techniques, from AI-powered market research and content creation to sophisticated customer service and analytics. Now, it is time to revisit the concept of an AI toolkit, but with a more advanced lens, diving deeper into the core tools that can become indispensable assets and introducing more sophisticated strategies for their use, including a focus on how an AI agent like Manus can assist in automating complex marketing workflows.

This chapter will provide expanded insights into some of the key AI tools previously mentioned (like ChatGPT, Claude, Perplexity, and Canva AI), focusing on their evolving capabilities and advanced applications for pickleball businesses. We will then dedicate a section to AI agents like Manus, detailing their unique value proposition for automating complex marketing workflows. Furthermore, we will delve into advanced prompt engineering techniques to help you elicit even more powerful responses from LLMs, explore how to integrate multiple AI tools for seamless workflows, and touch upon discovering niche AI solutions tailored for specific pickleball business needs.

Deeper Dives into Core AI Tools

The AI landscape is incredibly dynamic, with tools constantly updating and adding new features. Here is a look at some core tools with an emphasis on their advanced utility for pickleball marketing.

ChatGPT (OpenAI) & Claude (Anthropic) - Advanced Conversational AI

Recap of Strengths: Versatile text generation, brainstorming, coding assistance (ChatGPT), nuanced long-form content, strong reasoning, safety focus (Claude).

Advanced Applications for Pickleball Businesses:

Strategic Document Creation: Beyond blog posts, use them to draft initial versions of comprehensive marketing plans, business proposals for B2B pickleball services, or detailed standard operating procedures (SOPs) for marketing tasks.

Prompt Example (Claude for a Marketing Plan Outline): "Act as a seasoned CMO for a direct-to-consumer pickleball paddle brand. We are launching a new sustainable paddle line. Generate a detailed marketing plan outline for the first six months post-launch. Include sections for: Executive Summary, Target Audience Deep Dive, Product Positioning, SMART Goals, Key Marketing Channels (Digital & Offline), Content Strategy Pillars, Promotional Calendar, Budget Allocation (Conceptual), and Key Performance Indicators (KPIs). For each section, list three to five key sub-points or questions to address."

Advanced Data Analysis (with Code Interpreter/Plugins): If these tools have robust data analysis capabilities, use them to analyze sales data, customer feedback spreadsheets, or website analytics exports to uncover trends and insights without needing to be a data scientist.

Custom GPTs/Assistants: Develop custom versions of these AIs trained on your specific pickleball business's data, brand voice, and internal knowledge base.

Multi-Modal Capabilities: Leverage their ability to understand and generate content based on images or even voice inputs for richer marketing materials.

Perplexity AI - The Insight Engine

Recap of Strengths: Real-time, web-connected answers with sources, ideal for research and fact-checking.

Advanced Applications for Pickleball Businesses:

In-Depth Competitor Intelligence: Use Perplexity to find recent news, financial reports (if public), customer sentiment expressed in forums, and expert opinions about your competitors' strategies or new product launches in the pickleball market.

Legislative and Regulatory Monitoring: For pickleball businesses affected by local zoning laws or product safety standards, Perplexity can help track relevant updates.

Identifying Emerging Technologies or Trends: Use it to research broader trends in sports technology, AI in marketing, or sustainable business practices that could be adapted for your pickleball venture.

Prompt Example (Perplexity): "What are the latest advancements in AI-powered video analytics for sports coaching, and how could these be applied to pickleball training programs? Cite recent research papers or industry articles. Focus on research or articles published within the past 2 years. Organize the response into two parts: 1) Overview of advancements, and 2) Specific applications to pickleball."

Canva AI - Visual Content Powerhouse

Recap of Strengths: User-friendly design, AI-powered features like Magic Media, Magic Write, Magic Design.

Advanced Applications for Pickleball Businesses:

Brand Kit Consistency with AI: Utilize AI features that help maintain strict brand consistency across all visuals.

Bulk Creation of Social Media Assets: If Canva AI evolves to support batch processing, use it to generate multiple variations of a social media post from a single template and set of inputs.

Interactive Presentations and Pitch Decks: Leverage AI to help design more engaging and interactive presentations for B2B sales pitches.

AI Agents like Manus - Autonomous Assistance for Pickleball Businesses

We have mentioned AI agents like Manus throughout this book. Now, let us consolidate their unique value proposition.

Unique Value Proposition: An AI agent like Manus is designed to automate multi-step business tasks and complex workflows with minimal human oversight. It excels at integrating research, analysis, and action, effectively acting as an AI project manager or a digital operations specialist.

Strengths for Pickleball Businesses: Time savings, efficiency gains, enhanced data collection & synthesis, proactive operations, and scalability.

Ideal Use Cases (Recap & Expansion): Comprehensive competitor monitoring, automated market research, multi-channel lead nurturing, content lifecycle management, reputation management, and cross-platform analytics reporting.

Tasking an AI Agent: Unlike prompting an LLM for a single output, tasking an agent like Manus involves defining a strategic objective and often a series of steps or rules for it to follow. Clarity on the end goal, information sources, tools it might use, desired output format, and task frequency are crucial.

Advanced Example Task for Manus (New Product Opportunity Monitoring):

"`**Objective:** Identify potential new product opportunities or feature enhancements for our line of premium pickleball paddles by monitoring online discussions among advanced players.

Frequency: Continuous monitoring, with a summary report delivered every Friday.

Sources to Monitor: Reddit (r/Pickleball, r/AdvancedPickleball), PickleballForum.com (equipment sections), specific Facebook Groups (if accessible).

Keywords/Concepts to Track: Mentions of "wish my paddle had," "looking for a paddle that," "problem with [paddle type]," "new paddle technology," unmet needs related to spin, control, power, durability.

Actions for Manus: Scan sources daily, extract relevant text snippets, categorize snippets (e.g., "Desire for More Spin"), summarize top emerging themes in a weekly report with illustrative quotes and source links.`"

Advanced Prompt Engineering Techniques for Pickleball Marketing

Getting the most out of LLMs often comes down to how you prompt them. Advanced prompting can unlock more creative, accurate, and tailored outputs.

Role-Playing (Persona Prompting): Instruct the AI to adopt a specific persona (e.g., "Act as a marketing strategist"). Be specific about expertise and viewpoint.

Chain-of-Thought (CoT) Prompting / Step-by-Step Thinking: For complex tasks, ask the AI to "think step by step" or explain its reasoning.

Few-Shot Prompting (Providing Examples): Give the AI examples of the desired input-output format.

Zero-Shot Chain-of-Thought: Simply appending "Let's think step by step" can improve reasoning for complex queries.

Refinement through Iteration (Conversational Prompting): Treat it as a conversation. Ask for revisions (e.g., "Make this more concise," "Rewrite for a beginner").

Structured Output Formatting: Ask the AI to provide output in a specific format (table, JSON, Markdown).

Prompt Example: "Analyze customer reviews for our clinic. Extract name, main positive point, and main negative point. Present in a Markdown table: Name, Positive Feedback, Negative Feedback."

Negative Prompts (Specifying What NOT to Do): Tell the AI what to avoid (e.g., "Write a product description. Do not use marketing hype.").

Integrating Multiple AI Tools for Complex Workflows

The true power of AI in marketing often comes from combining the strengths of different tools. An AI agent like Manus can facilitate this, but even without one, you can manually create powerful toolchains.

Example Workflow: Launching a New Pickleball Clinic

Market Research (Perplexity AI): Research demand for specific clinic types locally.

Content Ideation (ChatGPT/Claude): Brainstorm clinic names, taglines, selling points.

Visual Creation (Canva AI): Design social media graphics, website banner, flyer.

Website Copy (ChatGPT/Claude): Draft landing page copy.

Email Announcement (ChatGPT/Claude): Write email to subscriber list

Social Media Scheduling (AI-Powered Social Media Tool): Schedule promotional posts.

Registration & Payment (Online Booking System): Use a system for easy online actions.

Automated Reminders (Email Platform or Manus): Set up reminders for participants.

Post-Clinic Feedback (AI Survey Tool & ChatGPT/Claude for analysis): Send survey, analyze responses.

Ongoing Monitoring & Reporting (Manus): Task Manus to track key metrics (registrations, feedback sentiment) and report regularly.

Discovering Niche AI Solutions for Pickleball

Beyond general-purpose AI tools, the market is seeing a rise in niche AI solutions tailored for specific industries or tasks. For the pickleball world, this might include:

AI for Pickleball Video Analysis: Tools that analyze game footage to provide players with insights on technique, shot selection, or strategy (potentially useful for coaches creating content or offering advanced services).

AI for Pickleball Facility Management: Software that uses AI to optimize court scheduling, manage memberships, or even control lighting and climate in a club.

AI-Powered Pickleball Equipment Recommenders: E-commerce sites might use AI to guide customers to the perfect paddle or shoes based on their playing style, skill level, and preferences.

Keep an eye on industry news, app stores, and technology forums for emerging AI tools that could give your pickleball business a unique edge. When evaluating niche tools, consider their integration capabilities, ease of use, cost, and specific relevance to your business challenges.

Your AI toolkit is not static; it is an evolving collection of resources. By staying curious, experimenting with new tools, mastering advanced prompting techniques, and strategically integrating different AI solutions—potentially orchestrated by an AI agent like Manus—you can continuously enhance your marketing effectiveness and drive sustainable growth for your pickleball business. The key is to view AI not just as a set of individual tools, but as a comprehensive, intelligent system that augments your capabilities and amplifies your strategic vision.

CHAPTER 12

The Future of AI in Pickleball Marketing

As we have journeyed through the applications of Artificial Intelligence in growing your pickleball business, it is clear that AI is not just a fleeting trend but a fundamental shift in how marketing is conceived, executed, and optimized. The tools and techniques discussed in this book represent the current state-of-the-art, but the field of AI is evolving at an exponential pace. Looking ahead, the capabilities of AI will only become more sophisticated, more integrated, and more indispensable. For pickleball entrepreneurs and marketers, understanding these future trajectories is key to staying ahead of the curve and continuing to innovate.

This chapter will explore the emerging AI trends and technologies poised to further transform pickleball marketing. We will discuss how to prepare your business for this next wave of AI and emphasize the importance of lifelong learning in this ever-changing, AI-driven landscape.

Emerging AI Trends and Technologies to Watch

The AI revolution is far from over. Several emerging trends are set to redefine the boundaries of what is possible in marketing:

Hyper-Personalization at an Unprecedented Scale: Expect AI to deliver truly individualized experiences across all touchpoints. Imagine websites that dynamically reconfigure their layout and content in real-time based on an individual visitor's profile and intent. AI could generate unique product descriptions or even custom-designed paddle graphics on the fly, tailored to a single user's preferences.

More Sophisticated and Autonomous AI Agents: Future AI agents will likely possess even greater autonomy, learning capabilities, and proactive decision-making. They might independently identify new marketing opportunities, negotiate ad placements, or even autonomously manage entire marketing campaign budgets, allowing marketers to focus almost entirely on high-level strategy and creativity.

Generative AI for Richer, More Immersive Content: Expect rapid advancements in AI's ability to create highly realistic and interactive 3D environments, virtual reality (VR) and augmented reality (AR) experiences, and complex, long-form video content. For pickleball, this could mean AI-generated virtual tours of a new club facility or AR filters that let users "try on" pickleball apparel virtually.

AI in Predictive Analytics for Proactive Marketing: Predictive capabilities will become more accurate and granular. AI might predict precisely when a pickleball player is likely to need new shoes based on their playing frequency and past purchase cycles, allowing for perfectly timed marketing messages.

Ethical AI and Explainable AI (XAI): As AI becomes more powerful, the demand for transparency, fairness, and accountability will intensify. Explainable AI (XAI) will become more important, allowing marketers to understand *why* an AI made a particular decision or recommendation.

AI-Powered Voice and Conversational Commerce: Conversational AI will become more natural and capable, enabling seamless voice-activated commerce. Pickleball players might book courts or order gear simply by talking to an AI assistant.

AI in the Metaverse and Web3 for Pickleball Communities: If the metaverse or Web3 concepts gain mainstream traction, AI will be integral in creating and managing virtual pickleball communities, AI-driven non-player characters (NPCs), unique digital collectibles, and personalized experiences within these digital realms.

Preparing Your Pickleball Business for the Next Wave of AI

Staying competitive in an AI-driven future requires a proactive approach:

Foster a Culture of Learning and Adaptability: Encourage yourself and your team to be curious about AI and open to adopting new tools and methods.

Invest in Data Infrastructure: High-quality, well-organized data is the fuel for most AI systems.

Start Small, Scale Smart: Begin with AI applications that solve your most pressing business challenges or offer the clearest ROI.

Prioritize Ethical Considerations: Build ethical AI practices into your marketing strategy from the outset.

Focus on Augmenting Human Skills: View AI as a tool to empower your team and enhance human creativity and strategic thinking.

Network and Collaborate: Stay connected with other businesses experimenting with AI.

Experiment with Emerging Tools: Early adoption of effective new technologies can provide a significant competitive advantage.

Lifelong Learning: Staying Ahead in the AI-Driven Landscape

The most important takeaway for navigating the future of AI in pickleball marketing is the commitment to lifelong learning. Stay informed by following reputable AI news sources, joining AI marketing communities, attending webinars, taking online courses, experimenting hands-on, and reading books and industry reports.

The future of AI in pickleball marketing is incredibly exciting. It promises a world where marketing is more personalized, efficient, intelligent, and ultimately, more effective at connecting pickleball businesses with the passionate community they serve. By embracing these emerging technologies with a strategic mindset, a commitment to ethical practices, and a passion for continuous learning, you can ensure your pickleball business not only keeps pace but leads the way in this AI-powered future.

CONCLUSION

Your AI-Powered Future in Pickleball Starts Now

We have embarked on an extensive journey through the landscape of Artificial Intelligence, exploring its profound potential to revolutionize how you market and grow your pickleball business. From laying a strategic foundation with AI-driven market research and customer understanding, to deploying AI in core marketing functions like content creation, email marketing, and social media engagement, and finally, to leveraging AI for advanced applications in customer service, analytics, and sophisticated workflow automation, the message is clear: AI is no longer a futuristic fantasy but a tangible, accessible, and powerful ally for businesses of all sizes in the vibrant pickleball ecosystem.

Recap of Key Takeaways: Your AI Playbook

Throughout this book, we have aimed to demystify AI and provide you with a practical playbook. Let us briefly revisit the core principles and capabilities we have uncovered:

- **Strategic Foundation is Key:** AI is most effective when applied to a well-understood market and a clearly defined customer base (Chapters 1 and 2).

- **AI Augments Human Ingenuity:** The true power of AI lies in its ability to augment your skills, automate repetitive tasks, and provide data-driven insights, freeing you up for higher-level strategy and creativity (Chapter 3).
- **Content is King, and AI is its Royal Scribe & Designer:** AI tools can significantly accelerate and enhance your ability to brainstorm, write, and visually design compelling content (Chapter 5).
- **Personalization at Scale is Achievable:** From prospecting (Chapter 4) to email marketing (Chapter 6) and social media (Chapter 7), AI enables more personalized experiences.
- **Do Not Neglect Local Opportunities:** AI can bring new efficiencies to local marketing and direct mail (Chapter 8).
- **Customer Loyalty is Built on Excellent Service:** AI can power instant customer support and help build smarter loyalty programs (Chapter 9).
- **Data-Driven Decisions Lead to Optimized Results:** AI transforms raw data into actionable insights for continuous refinement (Chapter 10).
- **Master Your AI Toolkit:** Core tools like LLMs and design platforms are essential, and understanding how to integrate them, potentially with the help of AI agents for workflow automation, can provide a significant edge. Advanced prompt engineering is key (Chapter 11).
- **The Future is AI-Driven:** The AI landscape will continue to evolve rapidly. A commitment to lifelong

learning and ethical application is paramount (Chapter 12).

The Power of Prompts and Practical Application

A recurring theme has been the importance of effective prompting and practical application. The numerous examples provided are starting points. The real magic happens when you adapt these prompts to your unique pickleball business, your specific audience, and your distinct brand voice. Experiment, iterate, and learn from each interaction with these AI tools.

Final Encouragement and Call to Action: Serve Up Your AI Success!

The journey into leveraging AI for your pickleball business might seem daunting at first, but like learning a new shot on the court, it begins with understanding the fundamentals, practicing consistently, and gradually building your confidence and skill. Start with one or two areas where AI can make an immediate impact.

Do not aim for perfection from day one. Embrace the learning process. The AI tools themselves are constantly learning and improving, and so will you. The pickleball industry is characterized by passion, community, and innovation – a perfect environment for the thoughtful application of Artificial Intelligence.

Take these strategies, tools, and prompts, adapt them, innovate with them, and most importantly, use them to serve your customers better. The court is set, the AI is ready. It is your time to serve up success.

COMPANION RESOURCE KIT

AI Mastery for Pickleball Business Owners

for those reading print editions please use this QR Code to Access:

Download the RESOURCE KIT including EVERY PROMPT

listed by Chapter HERE:

Or Scan here to Download

PROMPTS BY CHAPTER

CHAPTER 1: *Understanding Your Pickleball Market with AI*

Identifying Your Niche in the Pickleball Ecosystem (B2C vs. B2B)

To illustrate how AI can assist in brainstorming niche opportunities, consider using a Large Language Model (LLM) like ChatGPT or Claude.

Prompt Example (for ChatGPT/Claude):

"Assume the role of an experienced market analyst specializing in the sports industry. My objective is to identify underserved or high-potential niche markets within the U.S. pickleball ecosystem, considering both B2C and B2B opportunities. Generate 10 distinct niche ideas. For each idea, provide: 1. A concise description of the niche. 2. The primary target audience (consumer or business type). 3. Potential unique selling propositions (USPs) for a business operating in this niche. 4. Key considerations or potential challenges associated with this niche."

AI Tools for Market Trend Analysis

Tool Spotlight: Perplexity AI

* **Strengths:** Provides up-to-date information with direct source citations, making it effective for quick research, fact-checking, and understanding current discussions. Its conversational interface allows for follow-up questions to explore topics in greater depth.

* **Ideal Use Cases for Pickleball Businesses:** Researching new equipment trends, understanding regional demand for pickleball facilities, tracking competitor announcements, and fact-checking claims before publishing content.

Prompt Example (for Perplexity AI):

"Summarize the top market trends in the U.S. pickleball paddle industry from the past six months — especially around materials (e.g., carbon fiber vs. fiberglass), price trends, and what players are saying about performance and preferences. Include links to recent articles, retail data, and player reviews if available."

Competitor Analysis with AI: Uncovering Strengths, Weaknesses, and Opportunities

Prompt Example (for ChatGPT/Claude, after providing URLs or text from a competitor's site):

"Analyze the provided website content for [Competitor's Pickleball Club Name]. Identify: 1. Their primary target audience as suggested by the website's language and offerings. 2. Their key services and amenities. 3. Their main selling propositions or differentiators. 4. The overall tone and brand voice of their communications. 5. Any apparent weaknesses or areas where my new pickleball club could offer a superior alternative."

For ongoing competitor monitoring, an autonomous AI agent such as Manus can be particularly useful. You could configure a task for Manus to regularly monitor the websites and social media feeds of your key competitors.

Example Task for Manus:

"On a weekly basis, monitor the websites ([URL1], [URL2], [URL3]) and Instagram profiles ([Profile1], [Profile2], [Profile3]) of

my top three local pickleball competitor clubs. Track and report any of the following changes or new information:

** New membership offers or pricing adjustments.*

** Announcements of new programs, clinics, or events.*

** Significant new facility upgrades or additions.*

** Special promotions or discounts being offered.*

** Highly engaging social media posts (based on likes and comments). Compile these findings into a concise weekly report, highlighting the most strategically important updates."*

Listening to the Community: AI for Social Listening and Insights

Prompt Example (for ChatGPT/Claude, after providing a collection of anonymized comments):

"Analyze the following set of public comments from a pickleball forum discussing the new 'BrandX CarbonGlide' paddle. Identify:

1.The top three most frequently mentioned positive attributes of the paddle. 2.The top three most frequently mentioned negative attributes or concerns.

3.The overall sentiment expressed (positive, negative, mixed).

4.Any suggestions for improvement mentioned by the users.

5.Emerging themes or topics of discussion related to this paddle."

Prompt Example (for Manus to monitor these forums and mentions):

"Set up continuous monitoring of public online communities, forums, and review sites where pickleball players discuss products. Track all mentions

of the following keywords: 'BrandX', 'CarbonGlide', 'carbon fiber paddle', and 'lightweight pickleball paddle.

For each week, generate a summary that includes:

-The top 5 frequently mentioned features (positive or negative) of the BrandX CarbonGlide paddle.

-Any urgent or recurring customer service issues or complaints.

-A list of frequently asked questions that could be addressed in our support materials or website.

-Suggestions or requests for product improvements.

-Emerging competitor mentions or comparisons.

-A sentiment breakdown with volume trends.

Include direct quotes where helpful and link to original posts when available. Output as a weekly report in plain language, designed for marketing and product teams."

CHAPTER 2: *Defining Your Ideal Pickleball Customer/Client with AI*

<u>Creating Detailed Customer Personas with AI</u>

Prompt Example (for ChatGPT/Claude):

"Act as a marketing strategist for a new upscale pickleball club situated in a suburban area with a diverse population including young professionals and active retirees. Our objective is to develop detailed customer personas to inform our marketing strategies. Based on general knowledge of these demographic groups and their potential interest in pickleball, generate two distinct customer personas:

Persona 1: Young Professional (Age 28-40) Persona 2: Active Retiree (Age 60-75)

For each persona, include: 1. A representative name and a brief physical description (e.g., "Energetic Emily, mid-30s, athletic build"). 2. Demographics: Age range, typical occupation, general income bracket, family status. 3. Pickleball Experience Level: (e.g., Beginner, Intermediate, Advanced, New to the sport). 4. Goals & Motivations for Playing Pickleball: (e.g., Fitness, social interaction, competition, stress relief, skill development). 5. Challenges & Pain Points Related to Pickleball: (e.g., Difficulty finding court time, lack of suitable playing partners, risk of injury, cost of equipment/ membership, perceived intimidation for beginners). 6. Information Sources & Channels They Trust: (e.g., Social media platforms, local community forums, word-of-mouth referrals, sports publications, specific websites). 7. Key Marketing Messages Likely to Resonate: (e.g., "Connect and compete in a vibrant community," "Stay active and social through pickleball," "Learn pickleball in a supportive and welcoming environment"). 8. Potential Objections to Joining Our Upscale Club: (e.g., Membership fees, perceived exclusivity, commute time).

Ensure each persona is vivid, relatable, and grounded in plausible characteristics."

AI-Driven Segmentation: Targeting the Right Audience Segments

Prompt Example (Conceptual, for an AI analysis tool or LLM with data analysis capabilities):

"Analyze the provided customer dataset for our online pickleball equipment store. The dataset includes purchase history (items bought, purchase frequency, average order value), email engagement metrics (open rates, click-through rates), and basic demographics (age, location, where available). Identify at least three distinct customer segments based on purchasing behavior and engagement levels. For each segment, describe: 1. Key characteristics and behaviors. 2. Potential marketing approaches tailored to this segment.

For instance, one segment might be identified as 'High-Value Enthusiasts' who frequently purchase premium paddles and consistently engage with email content, contrasted with 'Occasional Players' who infrequently buy entry-level equipment."

Understanding Customer Needs and Pain Points with AI Surveys and Feedback Analysis

Prompt Example (for ChatGPT/Claude):

"I'm developing a member survey for a suburban pickleball club to assess satisfaction and gather ideas for improvement. Please create five **open-ended** and five **multiple-choice** questions that cover: court availability, facility quality, program variety (e.g., leagues, clinics), social engagement opportunities, and communication effectiveness.

Ensure the questions are **neutral and inclusive**, designed to prompt honest feedback from players of all skill levels. Where relevant, suggest multiple-choice formats (e.g., Likert scale, yes/no, or ranking).

At the end, include 2 optional **demographic or play-level questions** to help segment responses during analysis."

Tool Spotlight: NotebookLM - AI-Powered Survey Analysis

***Strengths:** NotebookLM allows you to upload your own documents — including survey responses, spreadsheets, and reports — and then query them using natural language. It excels at summarizing open-ended feedback, identifying recurring themes, surfacing sentiment, and highlighting actionable suggestions, all without requiring manual tagging or data cleaning. It's especially useful for making qualitative survey results instantly usable for small teams.

***Ideal Use Cases:**

- Summarizing post-event or post-program feedback
- Understanding key reasons for membership churn or declining engagement
- Surfacing patterns in open-ended survey responses about facility quality or communication
- Identifying recurring feature requests or pain points for program improvements

Prompt Example (for NotebookLM, after providing a list of anonymized customer complaints):

"Analyze the following customer complaints collected from our pickleball club members. For each of the five key operational areas — court availability, facility cleanliness/ maintenance, program variety (e.g., leagues or clinics), staff communication, and overall member experience — do the following:

1) Identify the most commonly reported issues.

2) Summarize sentiment (positive, negative, mixed) related to each area.

3) Highlight any recurring suggestions or requests for improvement.

4)Flag any complaints that may represent urgent or reputational risk.

Present your findings as a structured summary with bullet points under each category."

Example Task for Manus:

"Monitor the following sources for customer feedback regarding our pickleball coaching services: [Feedback Email Address], [Specific Facebook Group Thread URL], [Google My Business Review Page URL]. Once a week, compile all new feedback. Analyze this feedback to identify: * Positive comments and testimonials (extract illustrative quotes). * Common themes in negative feedback or suggestions for improvement. * Any urgent issues that require immediate attention. Summarize these findings in a weekly report, including representative quotes and an overview of customer sentiment."

Chapter 3: *Developing Your AI-Infused Pickleball Marketing Strategy*

Prompt Example (for ChatGPT/Claude):

"I own a new online store that sells premium pickleball paddles. My general marketing goal is to increase sales and build brand awareness, but I need help turning that into specific SMART goals.

Act as a marketing consultant. Based on the fact that I'm targeting adult recreational pickleball players (ages 35–65), and I currently average 100 website visitors per day with a small budget focused on online ads and content creation, help me create **three distinct SMART marketing goals for Q3**.

For each goal, clearly define the Specific, Measurable, Achievable, Relevant, and Time-bound elements. Format each SMART goal in bullet points for easy reference."

Chapter 4: *AI for Prospecting and Lead Generation*

Identifying Potential Customers (B2C) and Clients (B2B) with AI Tools

Prompt Example (for ChatGPT/Claude, for B2B prospecting):

"Act as a market research assistant. I'm seeking potential B2B clients for my company, which sells custom-branded pickleball apparel to teams and clubs.

Identify 10 pickleball clubs or leagues in the state of [State Name] that meet the following criteria:

– Appear to have a large and active membership base OR regularly host tournaments or events (2023–2024)

– Show signs of recent activity (e.g., social media, updated website)

– **Do not** have an existing online store selling branded apparel (on-site or via third-party vendors)

For each organization, provide:

– Name of the club or league

– Website URL (if available)

– A brief note on what makes them a strong B2B lead (e.g., "hosts 5+ tournaments per year," "1,000+ members")"

Example Task for Manus (B2B Prospecting for a Court Resurfacing Company):

"Research and compile a list of 20 pickleball facilities (clubs, community centers, private courts) within a 100-mile radius of [Your City, State] that were established in 2019 or earlier.

Identify those whose **online reviews or website photos** suggest their courts may be due for resurfacing. Look for signs such as cracks, faded lines, uneven surfaces, or safety concerns mentioned in reviews or visible in imagery.

Prioritize facilities that host regular tournaments or leagues. For each, provide:

Facility Name

Address

Website URL

Contact Person or Department (e.g., Facility Manager or Operations Head)

Notes from reviews or site content regarding court condition

Deliver the list in a **CSV file**."

<u>AI-Powered Outreach Strategies (Personalization at Scale)</u>

Prompt Example (for ChatGPT/Claude, after researching a specific pickleball club):

"Draft a personalized outreach email to [Club Manager Name], manager of [Pickleball Club Name]. I represent a company offering AI-powered court scheduling software that streamlines booking and reduces administrative workload for busy clubs.

Based on their website (which notes a manual booking system and a recent large tournament), the email should:

5. Acknowledge the recent event positively.
6. Introduce the software and its main benefit: optimizing court usage and simplifying scheduling.
7. Gently mention the challenges of manual systems, especially as demand grows.
8. Propose a 15-minute demo call to explore how the software can help their club specifically.

Keep the message **personalized, concise, and benefit-oriented**. If helpful, include a brief note of social proof (e.g., "similar clubs are already seeing results")."

Specific AI Strategies & Prompts for Pickleball Business Scenarios

AI Workflow for this Membership Drive – 3 Prompts

Prompt Example (for Manus to Monitor groups and forums):

"Monitor public online forums and community platforms (such as Facebook Groups, Reddit, and Nextdoor) for posts within a 25-mile radius of [Your City, State] that include phrases like 'looking for pickleball lessons,' 'where to play pickleball in [City],' 'beginner pickleball,' or similar expressions of interest.

For each post that matches, extract and compile the following information into a Google Sheet or CSV:

– Name or username of the poster

– Date of the post

– Full text of the post

– Platform/source (e.g., Facebook, Reddit)

– Direct link to the post (if available)

Update the list weekly, and prioritize posts that show clear intent to join, learn, or play pickleball locally."

Prompt Example (for ChatGPT to draft friendly, non-intrusive messages):

"Draft a friendly, conversational message I can send to individuals who recently posted online about wanting to learn or play pickleball in [Your City]. The message should invite them to a free introductory clinic or open house at our club. Keep the tone warm and helpful, not salesy. Mention that spots are limited to encourage early response, and include a simple call-to-action (e.g., reply to RSVP or click a link to register)."

Prompt Example (for Manus to manage sending and tracking):

"Send the message below to individuals listed in our recent prospecting sheet who expressed interest in pickleball. Use their preferred contact method (email or messaging, if available). Track who opens the message, who responds, and who clicks the RSVP link. Update the outreach tracking sheet daily with status labels: 'Sent,' 'Opened,' 'Replied,' 'Clicked RSVP,' or 'No Response.'"

1. **B2B App Sales to Clubs (e.g., a new pickleball tournament management app):**

- **Prospecting:** Task Manus to research and identify 50 pickleball clubs or tournament organizers across the country who frequently host events but appear to be using manual or outdated systems for registration and bracketing (based on their websites or event pages).

- **Prompt Example (for Manus):**

"Research and compile a list of 50 pickleball clubs, facilities, or tournament organizers across the U.S. that regularly host tournaments, leagues, or competitive events. Focus on organizations that appear to be using manual or outdated systems for registration, scoring, or bracket management—such as Google Forms, printable PDFs, spreadsheets, or handwritten brackets (as seen on their websites or event pages).

For each organization, provide the following in a CSV file:

Club or organization name

Website URL

Event type(s) hosted (e.g., leagues, tournaments)

Evidence of outdated systems (e.g., screenshots, wording, or links to forms)

Contact person or department (e.g., Tournament Director, Events Coordinator)

Location (City, State)

Prioritize organizations with recurring events or large participant numbers. Include a column for notes on any pain points or inefficiencies observed during your research."

Outreach Prompt (for ChatGPT/Claude, to be used by Manus in an email sequence): "Subject: Streamline Your

[Club Name/Tournament Series Name] Events with AI-Powered Management

Dear [Contact Person],

I came across your impressive [mention specific tournament or event] and noted the vibrant pickleball community you are fostering.

Managing successful tournaments involves many operational details. Our new AI-enhanced platform, [App Name], is designed specifically for pickleball event organizers like you to simplify registrations, automate bracket creation, manage player communication seamlessly, and provide real-time scoring updates – saving significant administrative time.

Would a brief 15-minute virtual demonstration be helpful to show you how [App Name] could make your next event smoother and more professional?

Sincerely, [Your Name]"

Chapter 5: *Crafting Compelling Content with AI*

AI for Brainstorming and Idea Generation

Prompt Example (for ChatGPT/Claude):

"Act as a content strategist for a pickleball coaching business that helps intermediate players (3.0–4.0 level) improve their strategy and mental game. Our audience values actionable tips, deeper insight into gameplay decisions, and relatable stories of improvement.

Generate 20 original content ideas suitable for:

- Long-form blog posts (~1000 words)

• 30–60 second Instagram Reels (scriptable moments or visual tips)

• Thoughtful LinkedIn articles targeting serious recreational players

• A weekly newsletter tip section

For each idea, provide:

– A concise, catchy title

– A 1–2 sentence description of the content angle and which format it's best suited for

Ensure ideas are **distinct**, balance both **strategic and mental game challenges**, and reflect our tone: clear, encouraging, and experience-based."

Writing Engaging Website Copy and Product Descriptions with AI

Tool Spotlight: ChatGPT & Claude

* **Strengths:** Both are excellent for generating various forms of text. ChatGPT is often noted for its versatility and speed, while Claude is recognized for its nuanced writing, strong reasoning capabilities, and ability to handle longer, more complex prompts, making it particularly suitable for thoughtful or detailed copy.

* **Ideal Use Cases:** Drafting website pages, blog posts, email sequences, social media captions, and scripts.

Prompt Example (for ChatGPT/Claude - About Us Page):

"Draft the "About Us" page copy for a new e-commerce brand called **Pickleball Performance Pro**, which sells high-performance pickleball paddles, shoes, and accessories. Our customers are competitive and serious recreational players who care deeply about their game and their gear.

Our brand values are **innovation, quality**, and a deep understanding of what pickleball players truly need. The tone should be **authoritative yet passionate and inspiring**—we're not just selling gear; we're helping players perform at their best.

Please structure the copy with:

- A strong opening paragraph that draws in performance-minded players
- A clear statement of our mission or philosophy
- A short paragraph about our commitment to testing, curating, and sourcing only the best equipment
- A personal note that the founders are avid players themselves and built this store to meet a gap they saw in the market

Highlight what makes our store different, such as expert guidance, premium selection, and a shared passion for the sport. Aim for approximately 400 words."

Prompt Example (for ChatGPT/Claude - Pickleball Paddle Description):

"Write a compelling and detailed product description for a new carbon fiber pickleball paddle named the"ShadowStrike X1." The target audience is intermediate to advanced players seeking a balance of power and control.

Key Features: * Face Material: T700 Raw Carbon Fiber with Grit Surface * Core: 16mm Reactive Polymer Honeycomb *

Weight: 7.9 - 8.2 oz (Mid-weight) * Grip: Cushioned, perforated, 4.25-inch circumference * Shape: Elongated with a slightly tapered head * Sweet Spot: Generous, enhanced by core technology

The description should: 1. Have a catchy headline. 2. Translate each feature into a clear player benefit (e.g., raw carbon for maximum spin, 16mm core for enhanced control and feel). 3. Emphasize the paddle's suitability for competitive play. 4. Use strong, evocative language. 5. Include a call to action (e.g., "Elevate your game with the ShadowStrike X1 today!"). Aim for 200-250 words."

Creating Eye-Catching Visuals with AI Design Tools

Tool Spotlight: Canva AI

* **Strengths:** Features an extremely user-friendly interface with a vast library of templates. Its AI capabilities (Magic Media for text-to-image, Magic Write for generating text within designs, Magic Design for instant template suggestions based on your content) significantly speed up the creative process for social media posts, flyers, presentations, and more.

* **Ideal Use Cases:** Creating social media graphics, email headers, website banners, simple marketing videos, presentations, and flyers for pickleball events or promotions.

Prompt Example (for Canva Magic Media - Text-to-Image):

"Create a vibrant, energetic image for an Instagram post announcing a weekend pickleball clinic. Show a diverse group of adults (ages 30–60) smiling and playing pickleball outdoors on a sunny day. Include modern, colorful athletic wear appropriate for warm weather. The style should be photorealistic with a slightly artistic, dynamic feel. Emphasize motion, fun, and the spirit of learning—some players should be mid-action while others are observing or laughing nearby."

Video Content Creation and Editing with AI

LLMs can draft scripts for short promotional videos, instructional clips, or social media reels.

Prompt Example (for ChatGPT/Claude - Short Video Script):

"Write a script for a 45-second Instagram Reel demonstrating the proper technique for a pickleball dink shot. The target audience is beginner to early intermediate players. The script should include: 1. A catchy hook/intro (5 seconds). 2. Clear, concise instructions for stance, grip, swing, and ball contact (30 seconds, broken into 2-3 visual cues). 3. A brief outro with a call to action (e.g., "Follow for more pickleball tips!") (10 seconds). The language should be encouraging and easy to understand. Indicate where visuals or demonstrations would occur."

Writing High-Converting Email Copy with AI

Prompt Example (for ChatGPT/Claude):

"Generate 10 compelling email subject lines for a weekly newsletter from a pickleball club. The newsletter typically includes club news, upcoming event announcements, a pickleball tip of the week, and member spotlights. Vary the tone – include options that are informative, exciting, and community-focused. Ensure some are short and punchy, and others are slightly more descriptive."

Whether it is a welcome email, a promotional offer, a newsletter, or an abandoned cart reminder, AI can draft the entire email.

Tool Spotlight: ChatGPT & Claude

* **Strengths:** Excellent for drafting various email types. Provide them with the goal of the email, the target segment, key information to include, and the desired tone.

Prompt Example (for Claude - Welcome Email Series):

"Act as an email marketing specialist for"PicklePro Gear," an online store selling premium pickleball equipment. Draft a three-part welcome email series for new newsletter subscribers. The goal is to introduce the brand, build trust, and encourage their first purchase.

Email 1 (Sent immediately after sign-up):

* Subject: Welcome to the PicklePro Gear Family! Your 10% Off Code Inside! * Content: Provide a warm welcome, briefly introduce PicklePro Gear (emphasizing passion for quality and a curated selection), offer a 10% discount code for their first

order, and set expectations for future newsletters (expert tips, new arrivals, exclusive deals).

Email 2 (Sent 2 days later): * Subject: Find Your Perfect Paddle with PicklePro Gear * Content: Highlight a key product category (e.g., paddles), offer advice on choosing the right paddle (linking to a relevant blog post or guide), and showcase two to three popular paddles with brief descriptions and benefits.

Email 3 (Sent 4 days later): * Subject: Don't Miss Out on These Pickleball Essentials! * Content: Showcase other essential gear (balls, bags, apparel), perhaps feature a customer testimonial, and gently remind them of their 10% discount code: If you haven't ordered yet, there's still time to use your 10% off code!

Maintain a friendly, helpful, and enthusiastic tone throughout the series. Use short paragraphs, optional bullet points for product highlights, and end each email with a clear CTA."

Prompt Example (for ChatGPT - Promotional Email):

"Draft a promotional email for a 48-hour flash sale on all pickleball shoes at "CourtSideKicks.com." The sale offers a 25% discount. The email should: 1. Have an urgent and exciting subject line. 2. Clearly state the offer (25% off all shoes) and the duration (48 hours only). 3. Highlight the benefits of upgrading pickleball shoes (e.g., better grip, support, injury prevention). 4. Showcase two to three popular shoe models with their original and sale prices. 5. Include a clear call-to-action button: "Shop the Sale Now!" 6. Create a sense of urgency to encourage immediate purchase. The tone should be energetic and persuasive."

AI for A/B Testing Subject Lines and Content

Prompt Example (for ChatGPT/Claude):

"Generate five alternative subject lines for A/B testing, each using a different psychological angle — one focused on **urgency**, one on **benefits**, one on **curiosity**, one on **scarcity**, and one on **social proof**. Keep the tone fun and relevant to recreational pickleball players."

Chapter 7: *Leveraging AI for Social Media Engagement*

Developing a Data-Driven Social Media Strategy with AI

Prompt Example (for ChatGPT/Claude, after providing links to competitor social media profiles):

"Analyze the Instagram presence of these three competitor pickleball apparel brands: [Profile URL 1], [Profile URL 2], [Profile URL 3]. Identify: 1. Common content themes or pillars they focus on. 2. Types of posts that generate the highest engagement (likes, comments, shares). 3. Their posting frequency and typical content formats (e.g., Reels, carousels, single images). 4. Their overall brand voice and visual style on Instagram. 5. Any apparent gaps or opportunities in their content strategy that my brand could leverage. Present your findings in a comparative table followed by a short strategic summary with actionable recommendations."

Prompt Example (for ChatGPT/Claude - Instagram Post for a Club):

"Our pickleball club, "The Dinking Dragons," just hosted a successful charity tournament. Draft three engaging Instagram caption options for a photo carousel showcasing highlights from the event (smiling players, action shots, check presentation). The tone should be celebratory, community-focused, and appreciative of participants and sponsors. Include relevant hashtags like #PickleballForGood #[ClubName] #[CityName]Pickleball #CharityEvent #PickleballCommunity. Each caption should be 1–2 short paragraphs max and include a natural call to action (e.g., 'Tag a teammate,' 'Thanks to everyone who made this happen,' etc.)."

Engaging Your Community with AI-Powered Tools

Prompt Example (for ChatGPT/Claude - Responding to a common question):

"A potential customer asked on our Facebook page: "What are your pickleball club membership rates for families?" Draft a friendly and informative response that: 1. Thanks them for their interest. 2. Briefly explains that we offer family packages. 3. Directs them to the membership page on our website for detailed pricing OR suggests they send a private message with their specific family size for a custom quote. 4. Invites further questions."

Running Effective AI-Targeted Social Media Ad Campaigns

Prompt Example (for ChatGPT/Claude - Facebook Ad Copy)

"Draft Facebook ad copy for a campaign promoting a new line of eco-friendly pickleball paddles from" GreenSwing Paddles." The target audience is environmentally conscious pickleball players aged 25-55.

Ad Creative Elements Needed:

* **Headline (maximum 40 characters):** Provide 3 variations.

* **Primary Text (maximum 125 characters):** Provide 2 variations (highlighting eco-friendly materials and performance).

* **Call to Action Button:** Suggest appropriate options (e.g., Shop Now, Learn More).

The ad should evoke a sense of sustainability, quality, and love for the game. Mention that a portion of profits goes to environmental charities."

Chapter 8: *Exploring AI in Direct Mail and Local Marketing*

AI for Hyper-Personalized Direct Mail Campaigns

Prompt Example (for ChatGPT/Claude - Personalized Postcard for a New Club):

"Act as a marketing copywriter for"[Your City] Pickleball Hub," a new indoor pickleball facility opening next month. We are sending a direct mail postcard to households within a 3-mile radius.

For a segment identified as "Families with children aged 8-16," draft postcard copy (approximately 75 words for one side) that: 1. Announces the grand opening. 2. Highlights family-friendly features (e.g., junior programs, family membership discounts, safe environment). 3. Includes a special offer for this segment (e.g., "First junior clinic free with family sign-up!"). 4. Provides a clear call to action (e.g., "Visit our website at [URL] or scan the QR code to learn more!"). The tone should be welcoming, energetic, and family oriented."

You could then create variations for other segments, such as "Young Professionals" (highlighting evening leagues and social events) or "Seniors" (emphasizing low-impact fitness and morning socials).

Optimizing Local SEO with AI for Pickleball Facilities and Stores

Prompt Example (for ChatGPT/Claude - GBP Description):

"Draft a Google Business Profile description (maximum 750 characters) for "Court Masters Pickleball Club" located in Springfield, IL. We offer eight indoor courts, a pro shop, lessons for all levels, leagues, and social events. Keywords to include are: "pickleball Springfield IL," "indoor pickleball courts," "pickleball lessons," and "pickleball leagues." The description should be inviting and highlight our key features."

Prompt Example (for ChatGPT/Claude - Localized Blog Post Idea):

"Generate a blog post title and outline for "Springfield Pickleball Pro Shop" targeting local customers. The post should focus on "Choosing the Right Pickleball Paddle for Illinois Weather Conditions (Considering Indoor and Outdoor Play)." Incorporate local relevance by mentioning common local playing environments or seasonal considerations in Springfield."

AI for Managing Online Reviews and Local Listings

Prompt Example (for ChatGPT/Claude - Responding to a Positive Review):

"A customer left the following 5-star review for our pickleball store, "The Paddle Rack": "Great selection and super helpful staff! Found the perfect paddle thanks to Sarah." Draft a warm and appreciative response to this review. Mention Sarah if appropriate and encourage the customer to visit again. Keep the tone professional and public-friendly, suitable for a Google Business review."

Prompt Example (for ChatGPT/Claude - Responding to a Negative Review):

"A customer left a 2-star review for our club: "Courts were too crowded, and the booking system was confusing." Draft a polite and constructive response that: 1. Thanks the customer for their feedback. 2. Acknowledges their concerns about crowding and the booking system. 3. Briefly mentions any steps being taken to address these issues (e.g., "We are working on optimizing court schedules and have a tutorial for the booking system on our website"). 4. Invites the customer to contact management directly to discuss their experience further. Avoid being defensive; aim for a resolution-oriented tone."

Chapter 9: *AI for Customer Service and Retention in Pickleball*

<u>Tool Spotlight: Various AI Chatbot Platforms</u>

* **Strengths:** Platforms such as Tidio, Intercom, Drift, or custom solutions using LLM APIs provide immediate responses, reduce wait times, free up human staff for more complex issues, and are available 24/7.

* **Example: Designing a Knowledge Base Snippet for a Pickleball Tournament Chatbot:** * **Question:** "How do I register for the upcoming Summer Slam tournament?"

* **Chatbot Answer (to be programmed):** "You can register for our Summer Slam tournament directly on our website! Just head to [Tournament Page URL] and click the 'Register Now' button. You will need to provide your name, email, playing level (e.g., 3.0, 3.5, 4.0+), and partner's name if playing doubles. Registration closes on [Date]. Let me know if you have any other questions!"

Personalizing Customer Communication with AI

Prompt Example (for ChatGPT/Claude - Membership Anniversary Email):

"Draft a warm and appreciative email to a pickleball club member, [Member Name], on their one-year membership anniversary with "The Pickleball Palace." The email should: 1. Congratulate them on their anniversary. 2. Thank them for being a valued member of the community. 3. Briefly highlight a positive aspect of the club (e.g., "We hope you have enjoyed the friendly games and improved your skills over the past year!"). 4. Offer a small token of appreciation (e.g., "As a thank you, enjoy a free guest pass on your next visit or 10% off in our pro shop this month!"). 5. Reinforce the club's commitment to providing a great pickleball experience. The tone should be personal and celebratory."

AI for Feedback Collection and Sentiment Analysis

Prompt Example (for ChatGPT/Claude, after providing customer service chat logs):

"Analyze the provided customer service chat logs from the past week for our pickleball equipment website. Identify: 1. The top three most common issues or questions raised by customers. 2. The overall sentiment of customer interactions (mostly positive, mixed, mostly negative). 3. Any recurring compliments about our service or products. 4. Any recurring complaints or suggestions for improving the customer experience. 5. Provide two to three direct (anonymized) quotes that illustrate key findings. Organize the output into a structured summary with bullet points and short subheadings for each section."

Chapter 10: *Measuring Success: AI for Marketing Analytics and Reporting*

AI Tools for Data Visualization and Reporting

Tool Spotlight: Google Looker Studio

* **Strengths:** Highly customizable for creating interactive dashboards that pull data from various Google services and other connectors. AI integration can enhance its ability to surface insights automatically.

* **Ideal Use Cases for Pickleball Businesses:** Creating a central marketing dashboard that shows website traffic, social media engagement, email campaign performance, and sales data (if integrated) all in one place.

Manus for Automated Multi-Source Reporting

One of the significant challenges in marketing analytics is consolidating data from disparate sources. An AI agent like Manus can be a powerful ally here.

Example Task for Manus (Weekly Marketing Performance Snapshot):

"`Every Monday morning, compile a Weekly Marketing Performance Snapshot for" The Pickleball Pro Shop." Gather the following data:

From Google Analytics 4: Total website users, new users, top three traffic sources, overall website conversion rate for newsletter sign-ups, and number of online paddle sales.

From Instagram Insights: Follower growth, average engagement rate per post, and reach of top three posts from the past week.

From Mailchimp (Email Marketing Platform): Open rate and click-through rate for the latest newsletter, and number of new subscribers.

From Shopify (E-commerce Platform): Total online sales revenue and average order value.

Present this data in a structured report with brief bullet-point summaries highlighting key changes or trends compared to the previous week. Identify any significant anomalies (e.g., a sudden 50% drop in website traffic from organic search)."

Chapter 11: *The Pickleball AI Toolkit Revisited (Expanded & Advanced)*

<u>Deeper Dives into Core AI Tools</u>

ChatGPT (OpenAI) & Claude (Anthropic) - Advanced Conversational AI

Recap of Strengths: Versatile text generation, brainstorming, coding assistance (ChatGPT), nuanced long-form content, strong reasoning, safety focus (Claude).

Advanced Applications for Pickleball Businesses:

Strategic Document Creation: Beyond blog posts, use them to draft initial versions of comprehensive marketing plans, business proposals for B2B pickleball services, or detailed standard operating procedures (SOPs) for marketing tasks.

Prompt Example (Claude for a Marketing Plan Outline): "Act as a seasoned CMO for a direct-to-consumer pickleball paddle brand. We are launching a new sustainable paddle line. Generate a detailed marketing plan outline for the first six months post-launch. Include sections for: Executive Summary, Target Audience Deep Dive, Product Positioning, SMART Goals, Key Marketing Channels (Digital & Offline), Content Strategy Pillars, Promotional Calendar, Budget Allocation (Conceptual), and Key Performance Indicators (KPIs). For each section, list three to five key sub-points or questions to address."

Advanced Data Analysis (with Code Interpreter/Plugins): If these tools have robust data analysis capabilities, use them to analyze sales data, customer feedback spreadsheets, or website analytics exports to uncover trends and insights without needing to be a data scientist.

Custom GPTs/Assistants: Develop custom versions of these AIs trained on your specific pickleball business's data, brand voice, and internal knowledge base.

Multi-Modal Capabilities: Leverage their ability to understand and generate content based on images or even voice inputs for richer marketing materials.

Perplexity AI - The Insight Engine

Recap of Strengths: Real-time, web-connected answers with sources, ideal for research and fact-checking.

Advanced Applications for Pickleball Businesses:

In-Depth Competitor Intelligence: Use Perplexity to find recent news, financial reports (if public), customer sentiment expressed in forums, and expert opinions about your competitors' strategies or new product launches in the pickleball market.

Legislative and Regulatory Monitoring: For pickleball businesses affected by local zoning laws or product safety standards, Perplexity can help track relevant updates.

Identifying Emerging Technologies or Trends: Use it to research broader trends in sports technology, AI in marketing, or sustainable business practices that could be adapted for your pickleball venture.

Prompt Example (Perplexity): "What are the latest advancements in AI-powered video analytics for sports coaching, and how could these be applied to pickleball training programs? Cite recent research papers or industry articles. Focus on research or articles published within the past 2 years. Organize the response into two parts: 1) Overview of advancements, and 2) Specific applications to pickleball."

Canva AI - Visual Content Powerhouse

Recap of Strengths: User-friendly design, AI-powered features like Magic Media, Magic Write, Magic Design.

Advanced Applications for Pickleball Businesses:

Brand Kit Consistency with AI: Utilize AI features that help maintain strict brand consistency across all visuals.

Bulk Creation of Social Media Assets: If Canva AI evolves to support batch processing, use it to generate multiple variations of a social media post from a single template and set of inputs.

Interactive Presentations and Pitch Decks: Leverage AI to help design more engaging and interactive presentations for B2B sales pitches.

AI Agents like Manus - Autonomous Assistance for Pickleball Businesses

We have mentioned AI agents like Manus throughout this book. Now, let us consolidate their unique value proposition.

Unique Value Proposition: An AI agent like Manus is designed to automate multi-step business tasks and complex workflows with minimal human oversight. It excels at integrating research, analysis, and action, effectively acting as an AI project manager or a digital operations specialist.

Strengths for Pickleball Businesses: Time savings, efficiency gains, enhanced data collection & synthesis, proactive operations, and scalability.

Ideal Use Cases (Recap & Expansion): Comprehensive competitor monitoring, automated market research, multi-channel lead nurturing, content lifecycle management, reputation management, and cross-platform analytics reporting.

Tasking an AI Agent: Unlike prompting an LLM for a single output, tasking an agent like Manus involves defining a strategic objective and often a series of steps or rules for it to follow. Clarity on the end goal, information sources, tools it might use, desired output format, and task frequency are crucial.

Advanced Example Task for Manus (New Product Opportunity Monitoring):

"`**Objective:** Identify potential new product opportunities or feature enhancements for our line of premium pickleball paddles by monitoring online discussions among advanced players.

Frequency: Continuous monitoring, with a summary report delivered every Friday.

Sources to Monitor: Reddit (r/Pickleball, r/AdvancedPickleball), PickleballForum.com (equipment sections), specific Facebook Groups (if accessible).

Keywords/Concepts to Track: Mentions of "wish my paddle had," "looking for a paddle that," "problem with [paddle type]," "new paddle technology," unmet needs related to spin, control, power, durability.

Actions for Manus: Scan sources daily, extract relevant text snippets, categorize snippets (e.g., "Desire for More Spin"), summarize top emerging themes in a weekly report with illustrative quotes and source links.`"

Advanced Prompt Engineering Techniques for Pickleball Marketing

Getting the most out of LLMs often comes down to how you prompt them. Advanced prompting can unlock more creative, accurate, and tailored outputs.

Role-Playing (Persona Prompting): Instruct the AI to adopt a specific persona (e.g., "Act as a marketing strategist"). Be specific about expertise and viewpoint.

Chain-of-Thought (CoT) Prompting / Step-by-Step Thinking: For complex tasks, ask the AI to "think step by step" or explain its reasoning.

Few-Shot Prompting (Providing Examples): Give the AI examples of the desired input-output format.

Zero-Shot Chain-of-Thought: Simply appending "Let's think step by step" can improve reasoning for complex queries.

Refinement through Iteration (Conversational Prompting): Treat it as a conversation. Ask for revisions (e.g., "Make this more concise," "Rewrite for a beginner").

Structured Output Formatting: Ask the AI to provide output in a specific format (table, JSON, Markdown).

- **Prompt Example:** "Analyze customer reviews for our clinic. Extract name, main positive point, and main negative point. Present in a Markdown table: Name, Positive Feedback, Negative Feedback."

Negative Prompts (Specifying What NOT to Do): Tell the AI what to avoid (e.g., "Write a product description. Do not use marketing hype.").

Integrating Multiple AI Tools for Complex Workflows

The true power of AI in marketing often comes from combining the strengths of different tools. An AI agent like Manus can facilitate this, but even without one, you can manually create powerful toolchains.

- **Example Workflow: Launching a New Pickleball Clinic**

  Market Research (Perplexity AI): Research demand for specific clinic types locally.

  Content Ideation (ChatGPT/Claude): Brainstorm clinic names, taglines, selling points.

  Visual Creation (Canva AI): Design social media graphics, website banner, flyer.

  Website Copy (ChatGPT/Claude): Draft landing page copy.

Email Announcement (ChatGPT/Claude): Write email to subscriber list.

Social Media Scheduling (AI-Powered Social Media Tool): Schedule promotional posts.

Registration & Payment (Online Booking System): Use a system for easy online actions.

Automated Reminders (Email Platform or Manus): Set up reminders for participants.

Post-Clinic Feedback (AI Survey Tool & ChatGPT/Claude for analysis): Send survey, analyze responses.

Ongoing Monitoring & Reporting (Manus): Task Manus to track key metrics (registrations, feedback sentiment) and report regularly.

Discovering Niche AI Solutions for Pickleball

Beyond general-purpose AI tools, the market is seeing a rise in niche AI solutions tailored for specific industries or tasks. For the pickleball world, this might include:

- **AI for Pickleball Video Analysis:** Tools that analyze game footage to provide players with insights on technique, shot selection, or strategy (potentially useful for coaches creating content or offering advanced services).
- **AI for Pickleball Facility Management:** Software that uses AI to optimize court scheduling, manage memberships, or even control lighting and climate in a club.

- **AI-Powered Pickleball Equipment Recommenders:** E-commerce sites might use AI to guide customers to the perfect paddle or shoes based on their playing style, skill level, and preferences.

GLOSSARY

AI and Marketing Terms

- **Artificial Intelligence (AI):** Computer systems capable of performing tasks that typically require human intelligence, such as learning, problem-solving, and decision-making.
- **Algorithm:** A set of rules or instructions given to an AI system or computer to help it learn or solve problems.
- **Autonomous AI Agent:** An AI system designed to operate independently to achieve goals, capable of planning, executing tasks, and interacting with its environment and other tools with minimal human oversight. An example discussed in this book is Manus.
- **A/B Testing (Split Testing):** A marketing experiment where two or more variants of a webpage, email, or ad are shown to different segments of an audience to see which one performs better.
- **B2B (Business-to-Business):** Businesses that sell products or services to other businesses.
- **B2C (Business-to-Consumer):** Businesses that sell products or services directly to individual consumers.

- **Brand Voice:** The unique personality and style a brand uses in its communications.
- **Chatbot:** An AI program designed to simulate human conversation, often used for customer service or information retrieval.
- **Click-Through Rate (CTR):** The percentage of people who see an ad or link and then click on it.
- **Content Marketing:** Creating and distributing valuable, relevant, and consistent content to attract and retain a clearly defined audience.
- **Conversion Rate:** The percentage of users or leads who complete a desired action (e.g., making a purchase, signing up for a newsletter).
- **CRM (Customer Relationship Management):** Software used to manage interactions and relationships with current and potential customers.
- **Customer Acquisition Cost (CAC):** The total cost associated with acquiring a new customer.
- **Customer Lifetime Value (CLV):** A prediction of the net profit attributed to the entire future relationship with a customer.
- **Customer Persona:** A semi-fictional, generalized representation of an ideal customer based on research and data.
- **Data Mining:** The process of discovering patterns in large data sets using methods at the intersection of machine learning, statistics, and database systems.

- **Direct Mail Marketing:** Sending physical promotional material (flyers, postcards, letters) to a target audience through postal mail.
- **Engagement Rate:** A metric that measures the level of interaction (likes, comments, shares, saves) that content receives relative to audience size.
- **Explainable AI (XAI):** AI systems that can explain their decisions or predictions in a way that humans can understand.
- **Generative AI (GenAI):** AI that can create new content, such as text, images, audio, video, or code.
- **Google Business Profile (GBP):** A free tool from Google that allows businesses to manage their online presence across Google, including Search and Maps.
- **Hyper-Personalization:** Tailoring marketing content, offers, and experiences to individual users at a highly granular level.
- **Key Performance Indicator (KPI):** A measurable value that demonstrates how effectively a company is achieving key business objectives.
- **Landing Page:** A standalone web page created specifically for a marketing or advertising campaign, where a visitor "lands" after clicking a link.
- **Large Language Model (LLM):** An AI model trained on vast amounts of text data to understand, generate, and manipulate human language (e.g., ChatGPT, Claude).
- **Lead Generation:** The process of attracting and converting strangers and prospects into someone who

has indicated interest in your company's product or service.

- **Lead Magnet:** A free item or service that is given away for the purpose of gathering contact details; for example, lead magnets can be trial subscriptions, samples, white papers, e-newsletters, and free consultations.
- **Lead Nurturing:** The process of developing relationships with buyers at every stage of the sales funnel, and through every step of the buyer's journey.
- **Lead Qualification:** The process of determining whether a prospect has the characteristics (e.g., budget, authority, need, timeline) to be a good fit as a customer.
- **Local SEO (Search Engine Optimization):** Optimizing your online presence to attract more business from relevant local searches.
- **Machine Learning (ML):** A subset of AI where systems automatically learn and improve from experience without being explicitly programmed.
- **Marketing Funnel:** A model illustrating the theoretical customer journey from initial awareness to purchase.
- **Metaverse:** A collective virtual shared space, created by the convergence of virtually enhanced physical reality and physically persistent virtual space.
- **Multimodal AI:** AI systems that can process and understand information from multiple types of data, such as text, images, audio, and video, simultaneously.

- **Natural Language Processing (NLP):** A branch of AI that helps computers understand, interpret, and generate human language.
- **Niche Market:** A small, specialized segment of a larger market.
- **Prompt Engineering:** The process of designing and refining input queries (prompts) to elicit desired or optimal responses from AI models, especially LLMs.
- **Return on Ad Spend (ROAS):** A marketing metric that measures the amount of revenue earned for every dollar spent on advertising.
- **Return on Investment (ROI):** A performance measure used to evaluate the efficiency or profitability of an investment.
- **Search Engine Optimization (SEO):** The process of improving the quality and quantity of website traffic to a website or a web page from search engines.
- **Segmentation:** Dividing a broad target market into subsets of consumers, businesses, or countries that have, or are perceived to have, common needs, interests, and priorities.
- **Sentiment Analysis:** The use of NLP, text analysis, and computational linguistics to identify, extract, quantify, and study affective states and subjective information.
- **SMART Goals:** Goals that are Specific, Measurable, Achievable, Relevant, and Time-bound.
- **Social Listening:** Monitoring social media conversations and mentions related to a brand, competitors, industry, or specific keywords.

- **Vanity Metrics:** Metrics that look good on paper but do not necessarily translate to meaningful business results.
- **Workflow Automation:** The design, execution, and automation of processes based on workflow rules where human tasks, data or files are routed between people or systems based on pre-defined business rules.

Recommended AI Tools (Pickleball Business Focus)

This list provides a starting point for exploring AI tools relevant to pickleball businesses. The AI landscape evolves rapidly, so always check for the latest features, pricing, and reviews before committing. (Prices/Tiers as of early 2025, subject to change).

1. Conversational AI & Text Generation:

ChatGPT (OpenAI)
Relevance: Content creation (blogs, social media, emails, scripts), brainstorming, coding help, summarization, drafting strategic documents.
Tiers: Free tier available; ChatGPT Plus (paid) offers access to more advanced models, faster responses, plugins, and features like DALL-E 3 image generation.
Pickleball Tip: Use custom instructions to maintain your brand voice for all pickleball marketing copy.
Claude (Anthropic)
Relevance: Long-form content, nuanced writing, professional communications, detailed analysis, ethical AI focus.
Tiers: Free tier with usage limits; paid upgrades for higher limits and priority access.
Pickleball Tip: Excellent for drafting detailed guides or thoughtful responses to complex customer inquiries.

Research & Real-Time Information:

Perplexity AI
Relevance: Real-time web-connected answer engine, research, fact-checking, market insights, competitor tracking. Provides sources.
Tiers: Free tier available; paid Pro version for more comprehensive search capabilities and features.
Pickleball Tip: Use it to quickly research competitor pickleball club activities or find recent statistics on pickleball growth trends.

Visual Design & Content Creation:
Canva AI (within Canva)
Relevance: User-friendly graphic design, text-to-image (Magic Media), content suggestions (Magic Write), quick design formatting (Magic Design).
Tiers: Robust free tier; paid Pro tier unlocks more AI features, templates, and stock assets.
Pickleball Tip: Perfect for creating eye-catching social media posts for your pickleball league, flyers for events, or email headers.
DALL-E 3 (via ChatGPT Plus or Bing Image Creator)
Relevance: High-quality image generation from text prompts for unique visuals.

Tiers: Accessible via ChatGPT Plus or free through Bing Image Creator (with limits).
Pickleball Tip: Generate custom illustrations for your pickleball blog or unique background images for your website.

Midjourney
Relevance: Advanced, artistic image generation from text prompts, known for highly detailed and stylized outputs.

Tiers: Paid subscription tiers.
Pickleball Tip: Use for creating distinctive brand imagery or conceptual art related to pickleball.

Autonomous AI Agents & Workflow Automation: * **Manus**
Relevance: Automating multi-step business tasks and complex marketing workflows, integrating research, analysis, and action with minimal oversight.
Tiers: Check for current access details (potentially invite/waitlist or tiered plans based on usage).
Pickleball Tip: Task Manus to monitor competitor pickleball paddle pricing weekly, compile social media engagement reports for your club, or manage lead nurturing sequences for potential B2B clients.

Productivity & Workspace AI:
Google Gemini (within Google Workspace)
Relevance: AI assistance within Google Docs, Sheets, Gmail, etc. Good for drafting emails, summarizing documents, simple data analysis if already in the Google ecosystem.

Tiers: Free version available; paid integration within Google Workspace plans.
Pickleball Tip: Use it in Gmail to quickly draft replies to common inquiries about your pickleball facility.

Notion AI (within Notion)
Relevance: AI-powered workspace for notes, documents, project management. Helps organize, summarize, and generate content within your business knowledge base.
Tiers: Limited free AI features in free Notion plans; paid add-on for more extensive use.

Pickleball Tip: Use it to summarize meeting notes from your pickleball club committee or draft marketing plans directly within your project management space.

Social Media Management with AI Features:
Buffer / Hootsuite / Sprout Social (and similar) *
Relevance: Many social media management platforms are incorporating AI for optimal posting times, content suggestions, hashtag recommendations, and basic analytics insights.
Tiers: Vary by platform, often with free limited tiers and multiple paid subscription levels.
Pickleball Tip: Use AI-suggested posting times to ensure your pickleball tournament announcements reach the maximum local audience.

AI for Video Creation/Editing (Emerging Category): *
Pictory.ai / InVideo AI / Descript
Relevance: Tools that use AI to create videos from scripts or blog posts, offer AI voiceovers, or simplify editing.
Tiers: Typically subscription-based, some may offer free trials or limited free versions.
Pickleball Tip: Turn a blog post about "Top 5 Pickleball Drills" into a simple animated video with AI voiceover for social media.

AI Chatbots for Websites:
Tidio / Drift / Intercom / ManyChat
Relevance: Engage website visitors, answer FAQs, qualify leads, and provide 24/7 customer support. *
Tiers: Most offer free basic tiers with limitations and various paid plans for more advanced features and higher volumes.
Pickleball Tip: Implement a chatbot on your pickleball club's website to answer common questions about membership, court bookings, and lesson schedules.

Disclaimer: This list is not exhaustive and inclusion does not imply endorsement. Always conduct your own research to find the tools that best fit your specific pickleball business needs and budget.

List of Prompts In Each Chapter

This appendix serves as a quick reference to the types of prompts discussed throughout the book. Refer to the specific chapters for more detailed context and variations.

Chapter 1: Understanding Your Pickleball Market with AI * Brainstorming Niche Opportunities * Researching Market Trends * Analyzing Competitor Website Content * Task for Ongoing Competitor Monitoring * Analyzing Online Community Comments

Chapter 2: Defining Your Ideal Pickleball Customer/ Client with AI * Generating Customer Personas * Analyzing Customer Data for Segments * Drafting Survey Questions * Analyzing Customer Complaints/Feedback * Task for Continuous Feedback Monitoring

Chapter 3: Developing Your AI-Infused Pickleball Marketing Strategy * Refining General Goals into SMART Goals

Chapter 4: AI for Prospecting and Lead Generation * Identifying B2B Client Types * Task for B2B Prospect Research * Drafting Personalized Outreach Emails * Script Outline for AI Chatbot Lead Qualification * Task for B2B App Sales Prospecting * Outreach Email Prompt for B2B App

Chapter 5: Crafting Compelling Content with AI * Generating Content Ideas for Various Formats * Drafting "About Us" Page Copy * Writing Product Descriptions *

For more information on Done-For-You Reports and intent based audiences, as well as information about how becoming a verified vendor of the Recreational Pickleball Players Association (RPPA) can help you access marketing tools contact us using the QR code below:

Made in the USA
Monee, IL
04 November 2026

67382323R00089